This report contains the collective views of an international group of experts and does not necessarily represent the decisions or the stated policy of the United Nations Environment Programme, the International Labour Organisation, or the World Health Organization.

Environmental Health Criteria 125

PLATINUM

Published under the joint sponsorship of the United Nations Environment Programme, the International Labour Organisation, and the World Health Organization

First draft prepared by Dr G. Rosner, Dr H.P. König, and Dr D. Coenen-Stass, Fraunhofer Institute of Toxicology and Aerosol Research, Germany

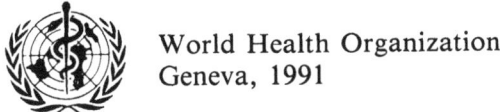

World Health Organization
Geneva, 1991

The **International Programme on Chemical Safety (IPCS)** is a joint venture of the United Nations Environment Programme, the International Labour Organisation, and the World Health Organization. The main objective of the IPCS is to carry out and disseminate evaluations of the effects of chemicals on human health and the quality of the environment. Supporting activities include the development of epidemiological, experimental laboratory, and risk-assessment methods that could produce internationally comparable results, and the development of manpower in the field of toxicology. Other activities carried out by the IPCS include the development of know-how for coping with chemical accidents, coordination of laboratory testing and epidemiological studies, and promotion of research on the mechanisms of the biological action of chemicals.

WHO Library Cataloguing in Publication Data

Platinum.

(Environmental health criteria ; 125)

1.Platinum - adverse effects 2.Platinum - toxicity 3.Environmental exposure I.Series

ISBN 92 4 157125 X   (LC Classification: QD 181.P8)
ISSN 0250-863X

©World Health Organization 1991

Publications of the World Health Organization enjoy copyright protection in accordance with the provisions of Protocol 2 of the Universal Copyright Convention. For rights of reproduction or translation of WHO publications, in part or *in toto*, application should be made to the Office of Publications, World Health Organization, Geneva, Switzerland. The World Health Organization welcomes such applications.

The designations employed and the presentation of the material in this publication do not imply the expression of any opinion whatsoever on the part of the Secretariat of the World Health Organization concerning the legal status of any country, territory, city, or area or of its authorities, or concerning the delimitation of its frontiers or boundaries.

The mention of specific companies or of certain manufacturers' products does not imply that they are endorsed or recommended by the World Health Organization in preference to others of a similar nature that are not mentioned. Errors and omissions excepted, the names of proprietary products are distinguished by initial capital letters.

Printed in Finland
91/8957 — Vammala — 5500

CONTENTS

ENVIRONMENTAL HEALTH CRITERIA FOR PLATINUM

1. SUMMARY AND CONCLUSIONS 13
   1.1 Identity, physical and chemical properties,
       analytical methods 13
   1.2 Sources of human and environmental exposure 14
   1.3 Environmental transport, distribution, and
       transformation 15
   1.4 Environmental levels and human exposure 16
   1.5 Kinetics and metabolism 17
   1.6 Effects on laboratory mammals and *in vitro*
       test systems 19
   1.7 Effects on humans 21
   1.8 Effects on other organisms in the laboratory
       and field 23

2. IDENTITY, PHYSICAL AND CHEMICAL PROPERTIES,
   ANALYTICAL METHODS 25
   2.1 Identity 25
   2.2 Physical and chemical properties 25
       2.2.1 Platinum metal 25
       2.2.2 Platinum compounds 25
   2.3 Conversion factors 30
   2.4 Analytical methods 30
       2.4.1 Sampling 30
       2.4.2 Sample pretreatment 30
       2.4.3 Detection and measurement 31
             2.4.3.1 Spectrophotometry 31
             2.4.3.2 Radiochemical methods 31
             2.4.3.3 X-ray fluorescence spectroscopy 35
             2.4.3.4 Electron spectroscopy for
                     chemical analysis 35
             2.4.3.5 Electrochemical analysis 35
             2.4.3.6 Proton-induced X-ray emission 36
             2.4.3.7 Liquid chromatography 36
             2.4.3.8 Atomic absorption spectrometry 37
             2.4.3.9 Inductively coupled plasma 38
             2.4.3.10 Inductively coupled plasma -
                      mass spectrometry 39

3. SOURCES OF HUMAN AND ENVIRONMENTAL
   EXPOSURE   41
   3.1 Natural occurrence   41
   3.2 Anthropogenic sources   41
       3.2.1 Production levels and processes   41
             3.2.1.1 World production figures   41
             3.2.1.2 Manufacturing processes   42
             3.2.1.3 Emissions from stationary
                     sources   43
             3.2.1.4 Emissions from automobile
                     catalysts   44
       3.2.2 Uses   46

4. ENVIRONMENTAL TRANSPORT, DISTRIBUTION,
   AND TRANSFORMATION   54
   4.1 Transport and distribution between media   54
   4.2 Biotransformation   56
   4.3 Ultimate fate following use   56

5. ENVIRONMENTAL LEVELS AND HUMAN EXPOSURE   58
   5.1 Environmental levels   58
       5.1.1 Ambient air   58
       5.1.2 Water and sediments   60
       5.1.3 Soil   62
       5.1.4 Food   63
       5.1.5 Terrestrial and aquatic organisms   63
   5.2 General population exposure   63
   5.3 Occupational exposure during manufacture,
       formulation or use   65

6. KINETICS AND METABOLISM   68

7. EFFECTS ON LABORATORY MAMMALS AND
   IN VITRO TEST SYSTEMS   75
   7.1 Single exposure   75
   7.2 Short-term exposure   76
   7.3 Skin and eye irritation; skin and respiratory
       sensitization   77
       7.3.1 Skin irritation   77
       7.3.2 Eye irritation   77
       7.3.3 Skin sensitization   79
       7.3.4 Skin and respiratory sensitization   79
       7.3.5 Respiratory sensitization   80
       7.3.6 Sensitization by other routes   80

	7.4	Reproductive toxicity, embryotoxicity, and teratogenicity	81
	7.5	Mutagenicity and related end-points	82
	7.6	Carcinogenicity and anticarcinogenicity	84
	7.7	Other special studies	85

		7.7.1 Effects on alveolar macrophages	85
		7.7.2 Non-allergic mediator release	86
		7.7.3 Effects on mitochondrial function	86
		7.7.4 Effects on the nervous system	86
		7.7.5 Side effects on cisplatin and its analogues	87
	7.8	Factors modifying toxicity	88
8.	EFFECTS ON HUMANS		89
	8.1	General population exposure	89
		8.1.1 Acute toxicity - poisoning	89
		8.1.2 Effects of exposure to platinum emitted from automobile catalysts	89
	8.2	Occupational exposure	89
		8.2.1 Case reports and cross-sectional studies	89
		8.2.2 Allergenicity of platinum and platinum compounds	92
		8.2.3 Clinical manifestations	92
		8.2.4 Immunological mechanism and diagnosis	93
		8.2.5 Predisposing factors	96
	8.3	Side effects of cisplatin	97
	8.4	Carcinogenicity	98
9.	EFFECTS ON OTHER ORGANISMS IN THE LABORATORY AND FIELD		99
	9.1	Microorganisms	99
	9.2	Aquatic organisms	99
		9.2.1 Plants	99
		9.2.2 Animals	100
	9.3	Terrestrial organisms	101
10.	EVALUATION OF HUMAN HEALTH RISKS AND EFFECTS ON THE ENVIRONMENT		104
	10.1	Evaluation of human health risks	104
		10.1.1 General population exposure	104
		10.1.1.1 Exposure	104
		10.1.1.2 Health effects	105
		10.1.2 Occupational groups	105
		10.1.2.1 Exposure	105

	10.1.2.2 Health effects	106
10.2	Evaluation of effects on the environment	107

11. RECOMMENDATIONS FOR PROTECTION OF
    HUMAN HEALTH AND THE ENVIRONMENT 110
    11.1 Pre-employment screening and medical
         evaluations 110
    11.2 Substitution with non-allergenic substances 110
    11.3 Employment screening and medical evaluations . 110
    11.4 Workplace hygiene 111

12. FURTHER RESEARCH 112

13. PREVIOUS EVALUATIONS BY
    INTERNATIONAL BODIES 114

    REFERENCES 115

    RESUME 140

    RESUMEN 154

WHO TASK GROUP ON ENVIRONMENTAL HEALTH CRITERIA FOR PLATINUM

Members

Dr V. Bencko, Institute of Hygiene, Charles University, Prague, Czechoslovakia

Dr R.E. Biagini, Division of Biomedical and Behavioral Sciences, National Institute for Occupational Safety & Health, Cincinnati, Ohio, USA *(Joint Rapporteur)*

Dr I. Farkas, National Institute of Hygiene, Budapest, Hungary

Dr U. Heinrich, Department of Environmental Hygiene, Fraunhofer Institute of Toxicology and Aerosol Research, Hanover, Germany

Dr R. Hertel, Fraunhofer Institute of Toxicology and Aerosol Research, Hanover, Germany

Professor G. Kazantzis, Centre for Environmental Technology, Royal School of Mines, London, United Kingdom

Professor A. Massoud, Department of Community, Environmental and Occupational Medicine, Faculty of Medicine, Ain Shams University, Cairo, Egypt *(Chairman)*

Dr R. Merget, Department of Internal Medicine, Hospital of the Johann Wolfgang Goethe University, Frankfurt am Main, Germany

Dr G. Rosner, Fraunhofer Institute of Toxicology and Aerosol Research, Hanover, Germany *(Joint Rapporteur)*

Dr A.E. Soyombo, Environmental & Occupational Health Division, Federal Ministry of Health, Lagos, Nigeria *(Vice-Chairman)*

Observers

Dr C.W. Bradford, Environmental, Health and Safety Services, Johnson Matthey Technology Centre, Reading, United Kingdom

Dr W.E. Mayr, Industrial Toxicology Department, Degussa AG, Hanau-Wolfgang, Germany

Secretariat

Dr P.G. Jenkins, International Programme on Chemical Safety, Division of Environmental Health, World Health Organization, Geneva, Switzerland

Dr E.M. Smith, International Programme on Chemical Safety, Division of Environmental Health, World Health Organization, Geneva, Switzerland

NOTE TO READERS OF THE CRITERIA DOCUMENTS

Every effort has been made to present information in the criteria documents as accurately as possible without unduly delaying their publication. In the interest of all users of the environmental health criteria documents, readers are kindly requested to communicate any errors that may have occurred to the Manager of the International Programme on Chemical Safety, World Health Organization, Geneva, Switzerland, in order that they may be included in corrigenda, which will appear in subsequent volumes.

* * *

A detailed data profile can be obtained from the International Register of Potentially Toxic Chemicals, Palais des Nations, 1211 Geneva 10, Switzerland (Telephone No. 7988400 or 7985850).

ENVIRONMENTAL HEALTH CRITERIA FOR PLATINUM

The WHO Task Group on Environmental Health Criteria for Platinum met in Rome, Italy, from 3 to 7 December 1990. Dr A. Mochi opened the meeting on behalf of the host country and Dr E. Smith welcomed the participants on behalf of the heads of the three IPCS cooperating organizations (UNEP/ILO/WHO). The Task Group reviewed and revised the draft monograph and made an evaluation of the risks for human health and the environment from exposure to platinum and certain platinum salts.

The first draft of this document was prepared by Dr G. Rosner, Dr H.P. König, and Dr D. Coenen-Stass, Fraunhofer Institute for Toxicology and Aerosol Research, Hanover, Germany. The second draft was prepared by Dr G. Rosner following circulation of the first draft to IPCS contact points. Particularly valuable comments on the draft were made by the European Chemical Industry Ecology and Toxicology Centre (ECETOC), the US Environmental Protection Agency, Food and Drug Administration, National Institute of Occupational Safety and Health, and Centers for Disease Control, the United Kingdom Department of Health, and the National Institute of Public Health, Norway. Dr C.W. Bradford gave valuable assistance in verifying the nomenclature of platinum compounds. Dr E.M. Smith and Dr P.G. Jenkins, both members of the IPCS Central Unit, were responsible for the overall scientific content and technical editing, respectively, of this monograph. The efforts of all who helped in the preparation and finalization of the document are gratefully acknowledged.

* * *

Financial support for the meeting was provided by the Ministry of the Environment of Italy. The Centro Italiano Studi e Indagini undertook the organization and provision of meeting facilities.

Partial financial support for the publication of this monograph was kindly provided by the United States Department of Health and Human Services, through a contract from the National Institute of Environmental Health Sciences, Research Triangle Park, North Carolina, USA - a WHO Collaborating Centre for Environmental Health Effects.

ABBREVIATIONS

AAS	atomic absorption spectrometry
BSA	bovine serum albumin
DC	direct current
DNA	deoxyribonucleic acid
ESCA	electron spectroscopy for chemical analysis
ETV	electrothermal vaporization
HSA	human serum albumin
ICP	inductively coupled plasma
Ig	immunoglobulin
LC	liquid chromatography
LC_{50}	median lethal concentration
MeB_{12}	methylcobalamin
MS	mass spectrometry
OVA	ovalbumin
PCA	passive cutaneous anaphylaxis
PGM	platinum-group metals
PIXE	proton-induced X-ray emission
PSH	platinum salt hypersensitivity
RAST	radioallergosorbent test
TLV	threshold limit value
TWA	time-weighted average
UV	ultraviolet

MOLECULAR FORMULAE OF PLATINUM COMPOUNDS

PtO	platinum(II) oxide
PtO_2	platinum(IV) oxide
$PtCl_2$	platinum(II) chloride
$PtCl_4$	platinum(IV) chloride
$Pt(NO_3)_2$	platinum(II) nitrate
$Pt(SO_4)_2$	platinum(IV) sulfate
$H_2[PtCl_4]$	hydrogen tetrachloroplatinate(II)
$H_2[PtCl_6]$	hydrogen hexachloroplatinate(IV) (commonly known as hexachloroplatinic acid)
$H_2[Pt(NO_2)_2SO_4]$	hydrogen dinitrosulfatoplatinate(II)
cis-$[PtCl_2(NH_3)_2]$	cis-diamminedichloroplatinum(II) (commonly known as cisplatin)
$trans$-$[PtCl_2(NH_3)_2]$	$trans$-diamminedichloroplatinum(II)
$[Pt(NH_3)_4]Cl_2$	tetraammineplatinum(II) chloride
$[Pt(NO_2)_2(NH_3)_2]$	diamminedinitroplatinum(II)
$[Pt(C_5H_7O_2)_2]$	bis(pentane-2,4-dionato)platinum(II) (commonly known as bis(acetylacetonato)platinum(II))
$[Pt\{(NH_2)_2CS\}_4]Cl_2$	tetrakis(thiourea)platinum(II) dichloride
$K_2[PtCl_4]$	potassium tetrachloroplatinate(II)
$K_2[PtCl_6]$	potassium hexachloroplatinate(IV)
$K_2[Pt(CN)_4]$	potassium tetracyanoplatinate(II)
$K[PtCl_3(NH_3)]$	potassium amminetrichloroplatinate(II)
$K_2[Pt(NO_2)_4]$	potassium tetranitroplatinate(II)
$Na_2[PtCl_4]$	sodium tetrachloroplatinate(II)
$Na_2[PtCl_6]$	sodium hexachloroplatinate(IV)
$Na_2[Pt(OH)_6]$	sodium hexahydroxyplatinate(IV)
$Na[Pt(NH_3)Cl_3]$	sodium amminetrichloroplatinate(II)
$(NH_4)_2[PtCl_4]$	ammonium tetrachloroplatinate(II)
$(NH_4)_2[PtCl_6]$	ammonium hexachloroplatinate(IV)
$Cs_2[Pt(NO_2)Cl_3]$	cesium trichloronitroplatinate(II)
$Cs_2[Pt(NO_2)_2Cl_2]$	cesium dichlorodinitroplatinate(II)
$Cs_2[Pt(NO_2)_3Cl]$	cesium chlorotrinitroplatinate(II)

1. SUMMARY

1.1 Identity, physical and chemical properties, analytical methods

Platinum (Pt) is a malleable, ductile, silvery-white noble metal with the atomic number 78 and an atomic weight of 195.09. It occurs naturally mainly as the isotopes ^{194}Pt (32.9%), ^{195}Pt (33.8%), and ^{196}Pt (25.3%). In platinum compounds the maximum oxidation state is +6, while the states +2 and +4 are the most stable.

The metal does not corrode in air at any temperature, but can be affected by halogens, cyanides, sulfur, molten sulfur compounds, heavy metals, and hydroxides. Digestion with aqua regia or Cl_2/HCl (concentrated hydrochloric acid through which chlorine is bubbled) produces hexachloroplatinic acid, $H_2[PtCl_6]$, an important platinum complex. When heated the ammonium salt of hexachloroplatinic acid produces a grey platinum sponge. A dispersive, black powder ("platinum black") results from reduction in aqueous solution.

The chemistry of platinum compounds in aqueous solution is dominated by the complex compounds. Many of the salts, particularly those with halogen- or nitrogen-donor ligands, are water-soluble. Platinum, like the other platinum-group metals, has a pronounced tendency to react with carbon compounds, especially alkenes and alkynes, forming Pt(II) coordination complexes.

There are various analytical methods for the determination of platinum. Atomic absorption spectrometry (AAS) and plasma emission spectroscopy provide high selectivity and specificity and are the method of choice for analysing platinum in biotic and environmental samples. With these methods detection limits of a few µg/kg or µg/litre have been obtained for various media.

Inductively coupled argon plasma atomic emission spectroscopy is superior to electrothermal AAS because of lower matrix effects and the possibility of simultaneous multi-element analysis.

Summary

1.2 Sources of human and environmental exposure

The average concentration of platinum in the lithosphere or rocky crust of the earth is estimated to be in the region of 0.001-0.005 mg/kg. Platinum is found either in the metallic form or in a number of mineral forms. Economically important sources exist in the Republic of South Africa and in the USSR. The platinum content of these deposits is 1-500 mg/kg. In Canada, platinum-group metals (platinum, palladium, iridium, osmium, rhodium, ruthenium) are found in copper-nickel sulfide ores at an average concentration of 0.3 mg/kg, but are concentrated to above 50 mg/kg during the refining of copper and nickel. Small amounts are mined in the USA, Ethiopia, the Philippines, and in Colombia.

World mine production of platinum-group metals, of which 40-50% is platinum, has steadily increased during the last two decades. In 1971, production was 127 tonnes (51-64 tonnes of platinum). Following the introduction of the automobile exhaust gas catalyst, world mine production of platinum-group metals increased to approximately 270 tonnes (108-135 tonnes of platinum) in 1987. In 1989, total platinum demand in the western world was approximately 97 tonnes.

The principal use of platinum derives from its exceptional catalytic properties. Further industrial applications relate to other outstanding properties, particularly resistance to chemical corrosion over a wide temperature range, high melting point, high mechanical strength, and good ductility. Platinum is also used in jewellery and dentistry.

Specific complexes of platinum, particularly *cis*-diamminedichloroplatinum(II) (cisplatin), are used therapeutically.[a]

---

[a] This monograph is specifically concerned with platinum and selected platinum compounds of occupational and/or environmental importance. A detailed discussion of the toxic effects of the anticancer drug cisplatin and its analogues in humans and animals is beyond the selected scope of the Environmental Health Criteria series as these substances are used primarily as therapeutic agents. In addition, their toxic properties are exceptional compared to those of other platinum compounds.

Data on emissions of platinum to the environment from industrial sources are not available. During the use of platinum-containing catalysts, some platinum may escape into the environment, depending on the type of catalyst. Of the stationary catalysts used in industry, only those used for ammonia oxidation emit significant amounts of platinum.

Automobile catalysts are mobile sources of platinum. According to limited data, platinum attrition from the old pellet-type catalyst is between 0.8 and 1.9 μg per km travelled. About 10% of the platinum is water-soluble.

With the new generation of monolith-type catalyst, results from engine test stand experiments with a three-way catalyst indicate that total platinum emission is lower by a factor of 100-1000 than in the case of pellet-type catalysts. At simulated speeds of 60, 100, and 140 km/h, total platinum emission was found to be between 3 and 39 ng/m^3 in the exhaust gas, corresponding to about 2-39 ng per km travelled. The mean aerodynamic diameter of emitted particles was between 4 and 9 μm in different test runs. There is limited evidence that most of the platinum emitted is in the form of the metal or surface-oxidized particles.

1.3 Environmental transport, distribution, and transformation

Platinum-group metals are rare in the environment, in comparison with other elements. In highly industrialized areas, elevated amounts of platinum can be found in river sediments. It is assumed that organic matter, e.g., humic and fulvic acids, binds platinum, aided perhaps by appropriate pH and redox potential conditions in the aquatic environment.

In soil, the mobility of platinum depends on the pH, redox potential, chloride concentrations of soil water, and the mode of occurrence of platinum in the primary rock. It is considered that platinum will be mobile only in extremely acid conditions or in soil water with a high chloride content.

In *in vitro* test systems it has been demonstrated that some platinum(IV) complexes, in the presence of plati-

Summary

num(II), can be methylated by bacterial methylcobalamin under abiotic conditions.

1.4 Environmental levels and human exposure

The data base concerning environmental concentrations is extremely limited due to the very low levels of platinum in the environment and the associated analytical problems.

Concentrations in ambient air samples taken near freeways in the USA before the introduction of the automobile catalyst were below the detection limit of 0.05 pg/m^3. Some recent data from Germany indicate that close to roads the platinum air concentrations (particulate samples) range from $\leq$ 1 pg/m^3 to 13 pg/m^3. In rural areas the concentrations were of a similar order of magnitude ($\leq$ 0.6 to 1.8 pg/m^3).

Ambient air concentrations of platinum close to roads resulting from the introduction of pellet-type automobile catalysts have been estimated on the basis of dispersion models and experimental emission data. Estimated platinum concentrations near and on roads ranged from 0.005 to 9 ng per m^3 for total platinum. As the total platinum emission from a monolith-type catalyst is lower, probably by a factor of 100 to 1000, than that of a pellet-type catalyst, the platinum concentrations for this type of catalyst would be in the picogram to femtogram per m^3 range.

In roadside dust deposited on broad-leaved plants at various sites in California, concentrations of 37-680 µg per kg dry weight were detected. Although the number of samples was limited, the results indicate that automotive catalysts release platinum to the roadside environment.

In plant chamber experiments, grass cultures exposed for four weeks to slightly diluted exhaust gas from an engine equipped with a three-way catalyst (simulated speed: 100 km/h) contained no platinum at a detection limit of 2 ng/g dry weight.

Investigations of the platinum concentrations in Lake Michigan sediments led to the conclusion that platinum has been deposited there over the past 50 years at a fairly uniform rate. Concentrations in sediment cores of 1 to 20 cm varied only between 0.3 and 0.43 µg/kg dry weight.

While no platinum levels have been reported for fresh waters, high concentrations (730 to 31 220 µg/kg dry weight) have been found in the sediments of a highly polluted cut-off channel of the Rhine river, Germany.

Samples of limber pines contained platinum levels ranging between non-detectable and 56 µg/kg (ash weight). However, the content of the adjacent soils was in the same range, and no accumulation tendency was indicated by these limited data.

In isolated samples of plants from an ultrabasic soil, platinum levels of 100-830 µg/kg (dry weight) were found.

Sea-water samples have been found to contain between 37 and 332 pg/litre. In sediment cores from the Eastern Pacific, platinum concentrations varied between 1.1 and 3 µg/kg (dry weight). The highest concentration (21.9 µg per kg) was found in offshore ocean sediments. In marine macroalgae, platinum concentrations of between 0.08 and 0.32 µg/kg dry weight have been found.

Blood platinum levels of 0.1 to 2.8 µg/litre have been found in the general population. In sera from occupationally exposed workers, levels of 150 to 440 µg per litre have been reported.

The data base for platinum concentrations at the workplace is limited. Due to analytical shortcomings, older data (0.9 to 1700 µg/m^3) are probably not reliable. However, from these data it can be assumed that exposure to platinum salts was higher than the occupational exposure limit of 2 µg/m^3 currently adopted by most countries. In recent workplace studies, concentrations either below the detection limit of 0.05 µg/m^3 or between 0.08 and 0.1 µg/m^3 have been measured.

1.5 Kinetics and metabolism

Following a single inhalation exposure (48 min) to different chemical forms of platinum (5-8 mg/m^3), most of the inhaled ^{191}Pt was rapidly cleared from the body. This was followed by a slower clearance phase during the remaining post-exposure period. Ten days after exposure to ^{191}PtCl$_4$, ^{191}Pt(SO$_4$)$_2$, ^{191}PtO$_2$, and ^{191}Pt metal, whole body

Summary

retention of ^{191}Pt was approximately 1, 5, 8, and 6%, respectively, of the initial body burden. Most of the ^{191}Pt that was cleared from the lungs by mucociliary action and swallowed was excreted via the faeces (half-time, 24 h). A small fraction of the ^{191}Pt was detected in the urine, indicating that very little was absorbed in the lungs and the gastrointestinal tract.

In a comparative study on the fate of ^{191}PtCl$_4$ in rats (25 µCi/animal) following different routes of exposure, retention was highest after intravenous administration, followed by intratracheal exposure. It was lowest after oral administration. Since only a minute amount of the ^{191}PtCl$_4$ given orally was absorbed, most of it passed through the gastrointestinal tract and was excreted via the faeces. After 3 days, less than 1% of the initial dose was detected in the whole body. Following intravenous administration, ^{191}Pt was excreted in almost equal quantities in both faeces and urine. Elimination was slower than after oral dosing. After 3 days whole body retention was about 65%, and after 28 days it was still 14% of the initial dose. For comparison, after these periods about 22% and 8%, respectively, were retained by the body following intratracheal administration.

Principal deposition sites are the kidneys, liver, spleen, and adrenals. The high amount of ^{191}Pt found in the kidney shows that once platinum is absorbed most of it accumulates in the kidney and is excreted in the urine. The lower level in the brain suggests that platinum ions cross the blood-brain barrier only to a limited extent.

In contrast to the water-soluble salts, the insoluble PtO$_2$ was only taken up in minute amounts even though the salt was administered in the diet at an extremely high level, which resulted in a total platinum consumption of 4308 mg per rat over the 4-week period.

For both the simple platinum salts and cisplatin, it has been established that there is an initial rapid clearance followed by a prolonged clearance phase during the remaining post-exposure period, and that there is no evidence for markedly different retention profiles. However, cisplatin is, due to high chloride concentrations suppressing hydration, very stable in extracellular fluids. This explains why it is excreted mainly in the unchanged form.

Its excretion, in contrast to that of the simple platinum salts, is primarily via the urine.

1.6 Effects on laboratory mammals and *in vitro* test systems

The acute toxicity of platinum depends mainly on the platinum species. Soluble platinum compounds are much more toxic than insoluble ones. For example, oral toxicity to rats (LD_{50} values) decreased in the following order: $Na_2[PtCl_6]$ (25-50 mg/kg) > $(NH_4)_2[PtCl_6]$ (195-200 mg/kg) > $PtCl_4$ (240 mg/kg) > $Pt(SO_4)_2 \cdot 4H_2O$ (1010 mg/kg) > $PtCl_2$ (> 2000 mg/kg) > PtO_2 (> 8000 mg/kg). For the two latter compounds no LD_{50} could be calculated.

In skin testing of albino rabbits, PtO_2, $PtCl_2$, $K_2[PtCl_4]$, $[Pt(NO_2)_2(NH_3)_2]$, $Pt(C_5H_7O_2)_2$ and *trans*-$[PtCl_2(NH_3)_2]$ were graded as non-irritant. $(NH_4)_2[PtCl_6]$, $(NH_4)_2[PtCl_4]$, $Na_2[PtCl_6]$, $Na_2[Pt(OH)_6]$, $K_2[Pt(CN)_4]$, $[Pt(NH_3)_4]Cl_2$, and *cis*-$[PtCl_2(NH_3)_2]$ appeared to be irritant, but to various degrees.

In eye irritation tests all tested platinum compounds showed irritating effects. *Trans*-$[PtCl_2(NH_3)_2]$ and $(NH_4)_2[PtCl_4]$ were found to be corrosive.

Intense breathing difficulties were observed after the intravenous injection of chloro-platinum complexes into guinea-pigs and rats, presumably due to non-allergic histamine release. This nonspecific histamine release has complicated the interpretation of both animal and human studies with respect to the diagnosis of allergic sensitization.

After subcutaneous and intravenous injection of $Pt(SO_4)_2$ three times a week for 4 weeks, there was no induction of an allergic state, as measured by skin tests (guinea-pigs and rabbits), passive transfer, and footpad tests (mice). Administration of platinum-egg-albumin complex also failed to sensitize the experimental animals.

Attempted sensitization of female hooded Lister rats with the free salt of ammonium tetrachloroplatinate, $(NH_4)_2[PtCl_4]$, applied via the intraperitoneal, intramuscular, intradermal, subcutaneous, intratracheal, and footpad routes, together with *Bordetella pertussis* adjuvant, was unsuccessful, as shown by the direct skin

Summary

test, passive cutaneous anaphylaxis (PCA) test or a radio-allergosorbent test (RAST). However, with platinum-protein conjugates positive PCA results have been reported.

In Cynomolgus monkeys *(Macaca fasicularis)* exposed to sodium hexachloroplatinate, $Na_2[PtCl_6]$, by nose-only inhalation at a level of 200 µg/m³, 4 h/day, biweekly for 12 weeks, significantly greater pulmonary deficits were observed by comparison with control animals. With exposure to ammonium hexachloroplatinate, $(NH_4)_2[PtCl_6]$, only concomitant exposure to ozone (2000 µg/m³) produced significant skin hypersensitivity and pulmonary hyper-reactivity.

In oral studies with male Sprague-Dawley rats, the salts $PtCl_4$ (182 mg/litre drinking-water) and $Pt(SO_4)_2 \cdot 4H_2O$ (248 mg/litre) did not affect normal weight gain within the observation period of 4 weeks. With a 3-fold increase in platinum concentration, weight gain was reduced by about 20% only during the first week, paralleling a 20% decrease in feed and water consumption.

Only limited experimental data are available for platinum effects on reproduction, embryotoxicity, and teratogenicity. $Pt(SO_4)_2$ (200 mg Pt/kg) caused reduced offspring weight in Swiss ICR mice from day 8 to 45 post-partum. The main effect of $Na_2[PtCl_6]$ (20 mg Pt/kg) was a reduced activity level of the offspring of mothers exposed on the 12th day of gestation. Solid platinum wire or foil is considered to be biologically inert and adverse effects following implantation into the uterus of rats and rabbits were probably due to the physical presence of a foreign object.

After intravenous administration of $^{191}PtCl_4$ to pregnant rats (25 µCi/animal) on day 18 of gestation, the placental barrier was crossed to a limited extent.

Several platinum compounds have been found to be mutagenic in a number of bacterial systems. In comparative studies cisplatin was several times more mutagenic than other tested platinum salts. In *in vitro* studies with mammalian cells (CHO-HGPT-system), the relative mutagenic activity of *cis*-$[PtCl_2(NH_3)_2]$, $K[PtCl_3(NH_3)]$, and $[Pt(NH_3)_3Cl]Cl$ was 100:9:0.3. The mutagenicity of $K_2[PtCl_4]$ and *trans*-$[PtCl_2(NH_3)_2]$ was marginal, whereas $[Pt(NH_3)_4]Cl_2$ was not mutagenic. No mutagenic activity was observed for the compounds $K_2[PtCl_4]$ and $[Pt(NH_3)_4]Cl_2$ in the *Drosophila melanogaster* sex-linked recessive lethal test, a mouse micronucleus test, and the Chinese hamster bone marrow test.

Except for cisplatin, no experimental data are available for the carcinogenicity of platinum and platinum compounds. For cisplatin there is sufficient evidence for carcinogenic effects on animals. However, cisplatin and its analogues are rather exceptional by comparison with other platinum compounds. This is reflected in the unique mechanism for their anti-tumour activity. Intrastrand DNA cross-links, formed only by the cis isomer at a certain position of guanine, are regarded as reasons for this anti-tumour activity. It appears that replication of DNA in cancer cells is impaired, while in normal cells the cisplatin lesions on guanine are repaired before replication.

1.7 Effects on humans

Exposure to platinum salts is mainly confined to occupational environments, primarily to platinum metal refineries and catalyst manufacture plants.

The compounds mainly responsible for platinum salt hypersensitivity[a] are hexachloroplatinic acid, $H_2[PtCl_6]$, and some chlorinated salts such as ammonium hexachloroplatinate, $(NH_4)_2[PtCl_6]$, potassium tetrachloroplatinate, $K_2[PtCl_4]$, potassium hexachloroplatinate, $K_2[PtCl_6]$, and sodium tetrachloroplatinate, $Na_2[PtCl_4]$. Complexes where there are no halogen ligands coordinated to platinum ("non-halogenated complexes"), such as $K_2[Pt(NO_2)_4]$, $[Pt(NH_3)_4]Cl_2$ and $[Pt\{(NH_2)_2CS\}_4]Cl_2$, and neutral complexes such as *cis*-$[PtCl_2(NH_3)_2]$, are not allergenic, since they probably do not react with proteins to form a complete antigen.

The signs and symptoms of hypersensitivity include urticaria, contact dermatitis of the skin, and respiratory

---

[a] The term "platinosis" is no longer used for platinum-salt-related disease, as it implies a chronic fibrosing lung disease such as silicosis. Instead, "platinum salt allergy", "allergy to platinum compounds containing reactive halogen ligands", and "platinum salt hypersensitivity" (PSH) have been used, the last being preferred.

Summary

disorders ranging from sneezing, shortness of breath, and cyanosis to severe asthma. The latency period from the first contact with platinum to the occurrence of the first symptoms varies from a few weeks to several years. Once sensitization is established, symptoms tend to become worse as long as the workers are exposed in the workplace but usually disappear on removal from exposure. However, if long-duration exposure occurs after sensitization, individuals may never become completely free of symptoms.

Although no unequivocal exposure concentration-effect relationship can be deduced from the available literature, the risk of developing platinum salt sensitivity seems to be correlated with exposure intensity. Metallic platinum seems to be non-allergenic. With the exception of one single reported case of an alleged contact dermatitis from a "platinum" ring, no allergic reactions have been reported.

The clinical manifestations of platinum salt hypersensitivity reflect a true allergic response. The mechanism appears to be a type I (IgE mediated) response. The possibility of IgE antibodies to platinum chloride complexes developing in sensitive people has been assumed on the grounds of *in vivo* and *in vitro* tests. It is believed that the platinum salts of low relative molecular mass act as haptens that combine with serum proteins to form the complete antigen.

Skin prick tests with dilute concentrations of soluble platinum complexes appear to provide reproducible, reliable, reasonably sensitive, and highly specific biological monitors of allergenicity. The compounds used for routine screening of exposed workers are $(NH_4)_2[PtCl_6]$, $Na_2[PtCl_6]$, and $Na_2[PtCl_4]$. The sensitivity and reliability of the skin prick test has not been achieved by any *in vitro* test available. In enzyme immunoassays and in radioallergosorbent tests (RAST), IgE antibodies specific to platinum chloride complexes have been found. Although a correlation with the results of prick tests was reported, the applicability of RAST for screening purposes was questioned because of its nonspecificity.

Only limited cross-reactivity between platinum and palladium salts has been found in skin testing and RAST. Reactions to the platinum-group metals other than platinum have only been seen in individuals sensitive to platinum salts.

Smoking, atopy, and nonspecific pulmonary hyperreactivity have been associated with platinum salt hypersensitivity and could be predisposing factors.

For the general population, there is a lack of data on the actual exposure situation in countries where the automobile catalyst has been introduced. The possible ambient air concentrations, estimated on the basis of a few emission data and dispersion models, are at least a factor of 10 000 lower than the occupational exposure limit value of 1 mg/m^3 adopted by some countries for platinum metal as total inhalable dust. Since the emitted platinum is most probably in the metallic form, the sensitizing potential of platinum emissions from automotive catalysts is probably very low. Even if part of the platinum emitted was soluble and potentially allergenic, the safety margin to the occupational exposure limit for soluble platinum salts (2 $\mu g/m^3$) would be at least 2000.

In a preliminary immunological study, extracts of particulate automobile exhaust samples were tested on three human volunteer subjects using a skin prick test. No positive response was elicited.

No data are available to assess the carcinogenic risk of platinum or its salts to humans. With regard to cisplatin, evidence for human carcinogenicity is considered inadequate.

1.8 Effects on other organisms in the laboratory and field

Simple complexes of platinum have bactericidal effects. The discovery that neutral complexes such as cisplatin selectively inhibit cell division without reducing cell growth of a variety of gram-positive, and especially, of gram-negative bacteria has led to their application in medicine as anti-tumour agents.

Growth and yield of the green alga *Euglena gracilis* were inhibited by the soluble hexachloroplatinic acid (250, 500, and 750 μg/litre) in a laboratory "microcosm". Cisplatin caused chlorosis and stunted growth in

Summary

the water hyacinth *Eichhornia crassipes* at a concentration of 2.5 mg/litre.

A 3-week exposure to hexachloroplatinic acid, $H_2[PtCl_6]$, resulted in an LC_{50} value of 520 µg Pt per litre in the invertebrate *Daphnia magna*. At concentrations of 14 and 82 µg/litre, reproduction, measured as total number of young, was impaired by 16 and 50%, respectively.

After short-term exposure to tetrachloroplatinic acid, $H_2[PtCl_4]$, in a static bioassay, 24-, 48-, and 96-h LC_{50} values of 15.5, 5.2, and 2.5 mg Pt/litre, respectively, were found for the coho salmon *(Oncorhynchus kisutch)*. General swimming activity and opercular movement were affected at 0.3 mg/litre. Lesions in the gills and the olfactory organ were noted at 0.3 mg/litre or more. Concentrations of 0.03 and 0.1 mg/litre had no effect.

There have been studies on the effects of platinum on terrestrial plants, all conducted with soluble platinum chlorides. The growth of beans and tomato plants in sand culture was inhibited by hexachloroplatinic acid at concentrations of 3×10^{-5} to 15×10^{-5} mol/kg (5.9-29.3 mg/kg). Of nine horticultural crops grown in hydroponic solution with platinum tetrachloride, $PtCl_4$ (0.057, 0.57, and 5.7 mg Pt/litre), dry weights were significantly reduced in tomato, bell pepper, and turnip tops, and in radish roots at the highest concentration. At this level, the buds and immature leaves of most species became chlorotic. In some of the species the low levels of $PtCl_4$ had a stimulatory effect on growth. In addition, transpiration was suppressed at the highest platinum concentration, probably due to increased stomatal resistance. Growth stimulation was also observed at low levels of platinum (0.5 mg Pt/litre), administered as potassium tetrachloroplatinate, $K_2[PtCl_4]$, in seedlings of the South African grass species *Setaria verticillata* grown in nutrient solution. After two weeks, the length of the longest roots had increased by 65%. At the highest concentration applied, i.e. 2.5 mg Pt/litre, phytotoxic effects were seen in the form of stunted root growth and chlorosis of the leaves.

2. IDENTITY, PHYSICAL AND CHEMICAL PROPERTIES, ANALYTICAL METHODS

2.1 Identity

Platinum is a malleable, ductile, silvery-white noble metal with the atomic number 78 and an atomic weight of 195.09. It occurs naturally mainly as the isotopes ^{194}Pt (32.9%), ^{195}Pt (33.8%), and ^{196}Pt (25.3%). In platinum compounds, the maximum oxidation state is +6, while the states +2 and +4 are the most stable.

The most important platinum compounds are listed in Table 1.

2.2 Physical and chemical properties

2.2.1 Platinum metal

The metal does not corrode in air at any temperature, but can be affected by halogens, cyanides, sulfur, molten sulfur compounds, heavy metals, and hydroxides. Digestion with aqua regia or Cl_2/HCl (concentrated hydrochloric acid through which chlorine gas is bubbled) leads to hexachloroplatinic acid, $H_2[PtCl_6]$, an important platinum complex.

Platinum has a coefficient of expansion almost equal to that of sodium-calcium-silicate glass and the two materials can be used in combination, e.g., in electrodes.

Some chemical and physical data on platinum and selected compounds are listed in Table 2.

2.2.2 Platinum compounds

The chemistry of platinum compounds in aqueous solution is dominated by the complex compounds. Many of the salts, particularly those with halogen- or nitrogen-donor ligands, are water-soluble. In biochemical processes, cis-trans effects in the quadratic coordination of platinum play an important role. Platinum, like the other platinum-group metals (PGM), has a marked tendency to react with

Table 1. Chemical names, synonyms, and formulae of elemental platinum and platinum compounds[a]

Chemical name	CAS registry number[b]	Synonyms	Formula
Element			
Platinum	7440-06-4		Pt
Binary compounds			
Platinum(II) chloride	10025-65-7	platinous chloride	$PtCl_2$
Platinum(IV) chloride	13454-96-1	platinum tetrachloride	$PtCl_4$
Platinum(II) oxide	n.a.	platinous oxide	PtO
Platinum(IV) oxide	1314-15-4	platinic oxide; platinum dioxide	PtO_2
Platinum sulfate	n.a.		$Pt(SO_4)_2 \cdot 4H_2O$
Platinum nitrate[c]	n.a.		$Pt(NO_3)_2$
Coordination complexes			
Hexachloroplatinic acid(IV)	16941-12-1	chloroplatinic acid; dihydrogen hexachloroplatinate	$H_2[PtCl_6]$
Sodium hexachloroplatinate(IV)	16923-58-3	disodium hexachloroplatinate; sodium chloroplatinate	$Na_2[PtCl_6]$
Potassium hexachloroplatinate(IV)	16921-30-5	potassium chloroplatinate; platinic potassium chloride	$K_2[PtCl_6]$
Potassium tetrachloroplatinate(II)	10025-99-7	platinum potassium chloride; potassium platinochloride	$K_2[PtCl_4]$
Ammonium tetrachloroplatinate(II)	13820-41-2	ammonium platinous chloride; ammonium chloroplatinite	$(NH_4)_2[PtCl_4]$
Ammonium hexachloroplatinate(IV)	16919-58-7	ammonium platinic chloride; ammonium chloroplatinate; "yellow salt"	$(NH_4)_2[PtCl_6]$
cis-Diamminedichloroplatinum(II)	15663-27-1	cisplatin; cis-platinum; DDP; CDDP; CPDD; CACP; CPCC; Peyron's chloride	cis-$[PtCl_2(NH_3)_2]$
trans-Diamminedichloroplatinum(II)	14913-33-8	trans-dichlorodiammineplatinum(II)	trans-$[Pt(NH_3)_2Cl_2]$

[a] From: Windholz (1976); Weast & Astle (1981)
[b] n.a. = not available
[c] Kral & Peter (1977)

Table 2. Physical and chemical properties of platinum and selected platinum compounds[a]

Chemical name	Relative atomic/molecular mass	Melting point[b] (°C)	Boiling point (°C)	Relative density (g/cm^3)	Crystalline form[c]	Solubility[d] Cold water	Solubility[d] Hot water	Other solvents
Platinum (Pt)	195.09	1772	3827 (± 100)	21.45^{20}	silver-metallic cubic cr.	ins	ins	sol aq. regia
Platinum(II) chloride (PtCl$_2$)	266.00	581[b] (in Cl$_2$)		6.05	olive-green, hexagonal cr.	sl sol		ins al, eth; sol HCl, NH$_4$OH
Platinum(IV) chloride (PtCl$_4$)	336.90	370[b] (in Cl$_2$)		4.303	brown-red cr.	v sol	v sol	sl sol, al, NH$_3$
Platinum(IV) oxide (PtO$_2$)	227.03	450		10.2	black powder	ins	ins	ins acid, aq. regia
Platinum(II) oxide (PtO)	211.09	550[b]		14.9	violet-black cr.	ins	ins	sol HCl; ins aq. regia
Platinum sulfate (Pt(SO$_4$)$_2$·4H$_2$O)	459.27				yellow plates	sol	dec	sol al, eth, acid
Hexachloroplatinic acid(IV) (H$_2$[PtCl$_6$]·6H$_2$O)	517.92	60		2.431	red-brown deliquescent cr.	v sol	v sol	sol al, eth
Sodium hexachloroplatinate(IV) (Na$_2$[PtCl$_6$])	453.77				yellow, hygroscopic cr.	sol		sol al

Table 2 (contd).

Chemical name	Relative atomic/molecular mass	Melting point[b] (°C)	Boiling point (°C)	Relative density (g/cm^3)	Crystalline form[c]	Solubility[d] Cold water	Hot water	Other solvents
Potassium hexachloroplatinate(IV) ($K_2[PtCl_6]$)	486.03			3.50	orange-yellow cr. or yellow powder	sl sol	sol	ins al
Potassium tetrachloroplatinate(II) ($K_2[PtCl_4]$)	415.26				ruby-red cr.	sol		
Ammonium tetrachloroplatinate(II) ($(NH_4)_2[PtCl_4]$)	373.00				dark ruby-red cr.	sol		
Ammonium hexachloroplatinate(IV) ($(NH_4)_2[PtCl_6]$)	443.91			3.06	orange-red cr. or yellow powder	v sol		ins al
cis-Diamminedichloroplatinum(II) (cis-$[PtCl_2](NH_3)_2$)	300.07	270[b]			orange cr.	sl sol[e]		
trans-Diamminedichloroplatinum(II) (trans-$[PtCl_2](NH_3)_2$)	300.07							

[a] Compiled from: Windholz (1976); Weast & Astle (1981); Neumüller (1987).
[b] dec = decomposes
[c] cr. = crystals
[d] al = alcohol (ethanol); dec = decomposes; eth = ether; ins = insoluble; sl = slightly; sol = soluble; v = very
[e] Tobe & Khokhar (1977)

carbon compounds, especially alkenes and alkynes, forming Pt(II) coordination complexes.

Platinum hexafluoride, PtF_6, has the highest oxidation state of the element and is a strong oxidizing agent; the noble gas xenon can be oxidized to XeF_2 and oxygen to O^{2+} (Hoppe, 1965).

Hexachloroplatinic acid, $H_2[PtCl_6]$, is formed by the reaction of platinum metal with aqua regia or Cl_2/HCl. When heated, the ammonium salt of this acid produces a grey platinum sponge. A black powder ("platinum black") is produced by reduction in aqueous solution. Depending on the pH value, hydroxides exchange the halogen ligands with OH^- in a stepwise manner, leading to $PtO_2 \cdot nH_2O$ after dehydration (n = 1, 2, 3, 4). Further heating gives rise to PtO at 400 °C, which decomposes to platinum and O_2 at 560 °C.

By heating hexachloroplatinic acid at 240 °C, $PtCl_2$ can be obtained. It has a hexameric structure (Pt_6Cl_{12}) in the solid state and is soluble in benzene. This compound forms $H_2[PtCl_4]$ in HCl.

Platinum forms a large number of Pt(II) and Pt(IV) complexes with the formulae:

Pt(IV): $[PtX_{6-n}(NH_3)_n]^{n-2}$ where n = 0-6; X = halogen ligand
Pt(II): $[PtX_{4-n}(NH_3)_n]^{n-2}$ where n = 0-4; X = halogen ligand

The chemical structures of two of the more important platinum complexes are shown below.

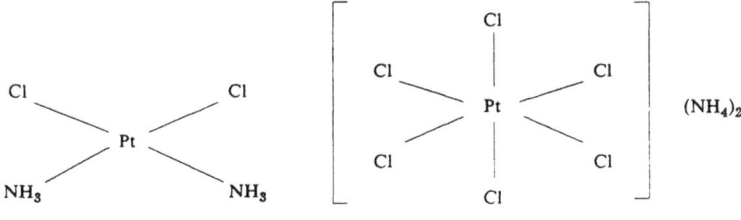

cis-diamminedichloroplatinum(II)   ammonium hexachloroplatinate(IV)

2.3 Conversion factors

Platinum   1 ppm = 7.98 mg/m^3
           1 mg/m^3 = 0.13 ppm

2.4 Analytical methods

2.4.1 Sampling

Samples of ores, minerals, and preconcentrated technical products can be obtained in a ground or powdered form. Metals and alloys can be collected as chips and shavings. Platinum on alumina pellets or monolithic supports must be comminuted before fusing or digesting (Potter & Lange, 1981). Electronic scrap may contain alloyed copper, nickel or lead. Melting with aluminium leads to a brittle alloy, which can be easily crushed to a powder.

Blood samples may be frozen and lyophilized (Pera & Harder, 1977), homogenized with substances like TRITON-X 100® (Priesner et al., 1981), and separated into plasma ultrafiltrate and proteins (Bannister et al., 1978) or, if appropriate, analysed directly without pretreatment.

With biological materials, homogeneous sampling is difficult and often requires destructive methods resulting in the loss of all information about the platinum species. Only the total content of platinum and its isotopes can be determined.

For the analysis of platinum in urine, the untreated original sample is usually unsuitable. Freeze-drying or a wet ashing procedure with subsequent reduction of volume is necessary for most analytical methods.

Other biological and environmental materials being investigated for very low levels of platinum need to be sampled in large amounts, with possible difficulty in homogenisation, digestion, storage, and matrix effects.

2.4.2 Sample pretreatment

Determination of total platinum content in some materials requires a digestion step, which is the prerequisite for enrichment and separation from other elements and organic substances. A modern wet digestion procedure (Knapp, 1985) avoids contact with materials other than quartz in order to reduce adsorption losses. In this way, organic matter is destroyed most effectively and contamination with platinum from other sources is minimized (Würfels et al., 1987).

In general, separation involves volatilization, distillation, lyophilization, extraction, coprecipitation, flotation, sorption, and other instrumental methods, such as electrodeposition, chromatographic separations, and thermal pretreatment in atomic absorption spectroscopy (AAS) procedures (Knapp, 1984).

A selection of extraction and sorption techniques is shown in Tables 3 and 4, respectively. For coprecipitation procedures, details can be found in the reports of Fryer & Kerrich (1978), Stockman (1983), Sighinolfi et al. (1984), Skogerboe et al. (1985), Amosse et al. (1986), and Bankovsky et al. (1987).

2.4.3 Detection and measurement

2.4.3.1 Spectrophotometry

Unless the native soluble platinum compounds have an inherent absorption spectrum, they can be treated with inorganic and organic reagents to form coloured, soluble complexes that can be measured by absorption spectrophotometry. Careful separation from other elements is important (see section 2.4.2). The detection limits achieved are in the low mg/kg (ppm) range (Jones et al., 1977; Brajter & Kozicka, 1979; Mojski & Kalinowski, 1980; Marone et al., 1981; Aneva et al., 1986; Puri et al., 1986).

2.4.3.2 Radiochemical methods

Neutron-activation analysis is a very sensitive method for determining submicrogram traces of platinum. It is at least one to several orders of magnitude more sensitive than the best of the spectrophotometric methods. For the determination of platinum a sensitivity of 1 ng absolute was estimated on irradiation of a sample for 1 month at a neutron flux of $10^{-2} cm^{-2}$-second, followed by a 2-h decay (NAS, 1977).

Table 3. Extraction procedures for separating platinum

Species	Matrix	Chemical modifier	Extraction medium	Elements separated	Reference
Pt(IV)	aqueous solutions	6 M HCl	isopentanol 4-methyl-2-pentanone	Al, Ca, Mg, Mn, Ni, Cr Cu, Pb (partially)	Aneva et al. (1986)
Pt(IV)	aqueous solutions	dithio-oxamide	tri-butyl phosphate	Ir(III), Rh(III)	Brajter & Kozicka (1979)
Pt(IV)	plant-processing solutions	S-(1-decyl)-N,N -diphenyl-isothiouronium bromide	variety of organic liquids	co-extraction of noble metals	Jones et al. (1977)
Pt(IV)	palladium(II) chloride	1,5-diphenylthio-carbazone	carbon tetra-chloride	Pd(II)	Marczenko & Kus (1987)
Pt(IV)	palladium metal	triphenylphosphine oxide	dichloroethane	Pd, Au	Mojski & Kalinowski (1980)
Pt(IV)	synthetic aqueous solutions	phenanthraquinone-monoxime	molten naphthalene	Fe, Cu, Ni, V, Cr, Al, Au, Ag Ir, Rh, Pd	Puri et al. (1986)
Pt(IV)	aqueous solutions	potassium butyl-xanthate	carbon tetra-chloride	-	Singh & Garg (1987)
Pt(IV)	automotive catalysts	bis-(2-furyl)-glyoxaldioxime	trichloro-methane	V, Mo, W	Wiele & Kuchenbecker (1974)

Table 3 (contd).

Species	Matrix	Chemical modifier	Extraction medium	Elements separated	Reference
Pt(II), Pt(IV)	synthetic aqueous solutions	1,4,7,10,13,16-hexa-azaoctadecane	4-methyl-2-pentanone	Fe(III)	Arpadjan et al. (1987)
Pt(II)	urine	Diethylammonium-diethyldithiocarbamate, NaSH	trichloro-methane	Ca, Zn, Fe(II) and Mn(II)	Borch et al. (1979)
Pt(II)	aqueous solutions	sodium diethyldithio-carbamate acetoni-trile, NaCl		co-extraction of Pd(II), Rh(II)	Mueller & Lovett (1987)
Pt(II)	plasma ultrafiltrate	sodium diethyldithio-carbamate methanol, H_2O		-	Andrews et al. (1984)
Pt	geological samples	sodium tetraborate, KCN	molten lead	-	Millard (1987)
Pt	geological samples	KCN, KOH	Ag, Au	co-extraction of noble metals	Le Houillier & De Blois (1986)
Pt	blood, hair, faeces, urine	HCl, $SnCl_2$	tri-n-octyl-amine, xylene	-	Tillery & Johnson (1975)
Pt	geological samples	sodium carbonate and sodium tetraborate	nickel sulfide	-	Robert et al., (1971)

Table 4. Sorption techniques for preconcentrating platinum

Species	Matrix	Sorption medium	Eluent	Elements separated	Reference
Pt	sea water	Bio-Rad Ag-1-X2	0.1 M HCl, 0.02 M thiourea	Ir	Goldberg et al. (1986); Hodge et al. (1986)
Pt	geological samples	Srafion NMRR	0.01 M HCl, 5% thiourea	high selectivity for transition metals	Kritsotakis & Tobschall (1985)
Pt	aqueous solutions	polyethenimine-methylthiourea suspended in water at pH 1		Co(II), Zn, Cd, In(III), Na	Geckeler et al. (1986)
Pt(II), Pt(IV)	aqueous solutions	Dowex 2X-8	75% NH_3 in H_2O	Au	Kahn & Van Loon (1978)
Pt(IV)	geological samples	Bio-Rad Ag-50W-X8	0.1 M HCl	-	Coombes & Chow (1979)
Pt(IV)	geological samples	P-TD	2 M $HClO_4$	Al, Mg, Cu, Fe, Ni, Cr	Grote & Kettrup (1987)
Pt(IV)	aqueous solutions	Hyphan	1 M $HClO_4$	Na, K, Cs, Mg, Ca, Al	Kenawy et al. (1987)
Pt(IV)	geological samples, soaps	Polyorgs	digestion $HClO_4$, H_2SO_4, HNO_3	coextraction noble metals	Myasoedova et al. (1985)
Pt(IV)	aqueous solutions	$(-CH_2-S-)_n$ (n≈1000)	6 M HCl	Co, Ni, Pb, Fe, Zn, Cd	Zolotov et al. (1983)

Radiochemical methods have been applied to the analysis of platinum in various matrices. The detection limits are 1-2 µg/kg in rock samples (Stockman, 1983), 30 µg per kg dry weight in plant material (Valente et al., 1982), 1-3 µg/kg dry weight (0.3 ng absolute) in plant material and animal tissue (Tjioe et al., 1984), and 100 µg/kg in airborne particulate matter (Schutyser et al., 1977).

2.4.3.3 X-ray fluorescence spectroscopy

This method permits the highly selective, sensitive, rapid, and non-destructive analysis of platinum. Zolotov et al. (1983) obtained a detection limit of 32 µg Pt per litre in aqueous solutions.

A new variant, total-reflection X-ray fluorescence spectrometry, has the advantage of small sample size (5 to 40 µg) with low absolute detection limits (Von Bohlen et al. 1987).

2.4.3.4 Electron spectroscopy for chemical analysis (ESCA)

ESCA is a technique typically applied in surface analysis involving a few surface atomic layers (1-2 nm). This technique is used for special purposes; for instance, Schlögl et al. (1987) analysed microparticles from automotive exhaust gas catalysts (see section 3.2.1.4).

2.4.3.5 Electrochemical analysis

Of the voltametric techniques available for element analysis, polarography, in particular, has been applied for the determination of platinum. Alexander et al. (1977a,b) described a pulse polarography method for the analysis of platinum in ores after fire-assay separation and preconcentration. By measuring the sensitive catalytic polarographic wave generated by the Pt(II)-ethylenediamine complex in alkali solutions a detection limit of 0.025 µg per kg was obtained. A similar technique was applied to the analysis of urine by Vrana et al. (1983), and the detection limit was 10 µg/litre.

However, these methods do not allow the direct determination of platinum in complex solutions due to interferences from some heavy metals and precipitation of platinum with other metals in the form of their hydroxides. In this respect, inverse voltametry is superior. Kritsotakis & Tobschall (1985) used the glassy carbon electrode for the determination of platinum traces in synthetic solutions. After preconcentration, 0.04 mg Pt/litre could be determined. This detection limit is sufficient for determining platinum in ores.

Using adsorptive cathodic stripping voltametry, Van den Berg & Jacinto (1988) analysed sea-water samples (see section 5.1.2). The detection limit was 7.8 pg Pt/litre.

Hoppstock et al. (1989) developed a sensitive voltametric method for determining platinum in the ng/kg range in biotic and environmental materials. The overall recovery of platinum was reported to be 97% or more.

Nygren et al. (1990) described an adsorptive voltametric method for the measurement of platinum in blood. The detection limit for a 100-μl sample was 0.017 μg per litre.

2.4.3.6 Proton-induced X-ray emission (PIXE)

PIXE requires only small sample sizes (1-10 mg), but is a time-consuming and labour-intensive method. Owing to the substantially lower background, the detection limits are lower by a factor of 1000 than for X-ray fluorescence methods. Methods for analysing water samples, air, and biological tissues have been described by Rickey et al. (1979), Wolfe (1979), and Thompson et al. (1981).

2.4.3.7 Liquid chromatography (LC)

Marsh et al. (1984) published an adsorption chromatography method in which the analyte was first separated with an ODS Hypersil® column, reacted with $NaHSO_3$, and then detected by UV absorption. The detection limit for cisplatin was 40-60 μg/litre. For the malonate derivates, Van der Vijgh et al. (1984) reported a detection limit of 300-1200 μg/litre for human body fluids.

Ebina et al. (1983) analysed Pt(II) in aqueous solutions that were modified with EDTA, ethanoic acid, and maleonitriledithiol. The spectrophotometric detection limit for this partition ion-pair method was 0.2 ng per litre.

Using an ion exchange chromatography method, Rocklin (1984) separated Pt(IV) as the hexachlorocomplex on a polar anion exchange column and determined the complex by UV. For samples digested in aqua regia, a detection limit of 30 µg/litre can be obtained without preconcentration and < 1 µg/litre after preconcentration.

1.3.8 Atomic absorption spectrometry (AAS)

AAS is a method of high selectivity and specificity and is often the method of choice in analysing platinum in biological and environmental samples. However, there are problems with background radiation deriving from molecules and radicals, especially from unseparated matrix. These interferences can be partly overcome by background compensation through a radiation continuum or by the application of the "Zeeman" effect. To determine platinum in the range of the detection limit, an accurate separation from matrix is essential.

For platinum determinations in biological materials, Farago & Parsons (1982) recommended wet digestion in nitric acid and the removal of residual nitrates by hydrochloric acid. Brown & Lee (1986) proposed totally pyrolytic cuvettes for graphite furnace AAS, thus achieving a greater sensitivity for refractory metals. These results were confirmed by Schlemmer & Welz (1986). Although platinum does not form a stable carbide, there was an effect on the wall material of the carbon rod. Electrographite tubes coated with pyrolytic graphite were found to be superior to glassy carbon tubes (Welz & Schlemmer, 1987).

LeRoy et al. (1977) described a method for the detection of platinum in biological samples that used controlled dehydration and ashing with rapid sample evaporation to detect low levels of platinum. This method did not suffer as much from matrix interference as other AAS graphite furnace methods. The method can be used to detect platinum down to approximately 30 µg/kg (30 ppb).

Hodge et al. (1986) determined platinum down to pg per litre levels in marine waters, sediments, and organisms. Sea water was extracted with an anion exchanger (Table 4), eluted, and purified by acid digestion. In a second step, platinum was obtained from the solution with an anion exchanger, stripped again from the bead, and injected. Using a similar technique, Hodge & Stallard (1986) determined platinum in roadside dust.

Jones (1976) digested urine and blood samples with nitric and perchloric acids. The samples were diluted after cooling and injected onto carbon rods. The minimum detectable platinum concentration in 5-g samples was 30 µg per litre.

McGahan & Tyczkowska (1987) dried and ashed tissues and fluids and diluted the residue with different acids before direct injection. The detection limits were 6 µg per kg or 6 µg/litre.

Bannister et al. (1978) separated protein-bound platinum and free circulating compounds by centrifugal ultrafiltration. In the ultrafiltrate, platinum compounds were chelated with ethylenediamine, extracted on a cation exchange paper disc, eluted, and injected. The minimum working concentration was 35 µg/litre of plasma.

Alt et al. (1988) described a simple and reliable method which included high-pressure ashing (cf. Knapp, 1984), separation by extraction, and detection by graphite furnace AAS. This method was recommended for analysing biological and other materials down to the µg/kg range.

König et al. (1989) determined platinum in the particulate emissions in engine test-stand experiments (see section 3.2.1.4) using a high-pressure digestion without a separation. The authors studied the matrix influences with respect to the concomitant elements and found interferences from Al, Pb, Ca, Zn, P and, most severely, from Si, but under the controlled test conditions no interference effects were observed. In particle-free condensates of automotive exhaust gas, a detection limit of 0.1 ng/ml was achieved by the method of signal addition described by Berndt et al. (1987).

2.4.3.9 Inductively coupled plasma (ICP)

The generation of plasmas is a further development of chemical flame methods. They have a wide temperature range, a transparency for the UV spectral lines, and are predominantly insensitive against interfering chemical reactions in the excitation zone that occur with chemical flames. Plasma excitation allows the determination of several elements simultaneously and is, because of minor matrix effects, easy to calibrate over many orders of magnitude. Two methods of generating a plasma are currently used: firstly with direct current (DC) and secondly with a high frequency current (20-80 MHz, inductively coupled plasma, ICP). The ICP method works with an argon plasma and temperatures of 4000-8000 K. Due to the increasing ionization effects, the aerosol feeding is controlled by cooling devices.

Boumans & Vrakking (1987) discussed standard values for a 50-MHz ICP, considering effects of source characteristics, noise, and spectral band-width, and obtained a detection limit for the platinum spectral line at 214.42 nm of 7.2 µg/litre.

Maessen et al. (1986) studied the influence of chloroform on the platinum signal at 203.65 nm. The detection limits by this method were affected by chloroform and ranged from 30-400 µg/litre.

Wemyss & Scott (1978) determined platinum-group metals and gold in ores after three different digestions. The method allowed determination down to 0.13 mg/litre for the 299.8-nm line.

Fox (1984) reported interferences from aluminium and magnesium in direct current methods. A buffer of lithium and lanthanum compounds suppressed this effect.

Lo et al. (1987) described a simple method for determining platinum in urine with a working range down to 50 µg/litre (50 ppb) under direct application of acidified samples. Electrothermal vaporization (ETV) was used for generating plasma-suitable aerosols by Matusiewicz & Barnes (1983). They determined platinum at the mg/litre level in human body fluids directly. A similar procedure was used by Belliveau et al. (1986).

4.3.10 Inductively coupled plasma - mass spectrometry (ICP-MS)

Combining ICP with a mass spectrometer has new advantages in analytical spectroscopy. Elemental ions generated from an aerosol or an electrothermal vaporization unit are separated by a quadrupole and detected as isotopes at low level. The ETV device allows determination down to the pg/ml range.

Thompson & Houk (1986) used an ion-pair reversed-phase liquid chromatography assay via a continuous flow ultrasonic nebulizer and an ICP torch with a mass spectrometer. In synthetic solutions detection limits of 7 µg/litre (7 ppb) were obtained.

Gregoire (1988) compared the results from the ICP-MS-ETV with neutron activation analysis and the ICP-MS solution nebulization method in the ng/ml concentration range and found good agreement.

For the analysis of air samples, the NIOSH Manual of Analytical Methods (Eller, 1984a) describes a method based on inductively coupled argon plasma atomic emission spectroscopy. The working range is 0.005-2.0 mg/m^3 with a 500-litre air sample. However, long sampling periods are required for measuring soluble platinum compounds in the workplace and the method does not distinguish between soluble and insoluble platinum. Similar methods are recommended for the analysis of platinum in blood and tissues (Eller, 1984b) and in urine (Eller, 1984c).

The method recommended by the United Kingdom Health and Safety Executive (1985) has a precision better than 8%, measured as a coefficient of variation, for samples of a minimum of 120 litres in the range 1-15 µg Pt/m^3. The sensitivity of this method can be improved by 100-1000 fold by using ICP-MS instead of carbon furnace atomic absorption spectrometry.

3. SOURCES OF HUMAN AND ENVIRONMENTAL EXPOSURE

3.1 Natural occurrence

The six platinum-group metals, platinum, palladium, rhodium, ruthenium, iridium, and osmium, were probably concentrated mainly in the iron-nickel core during the earth's formation. This explains their relatively low presence in the lithosphere (rocky crust) of the earth (Goldschmidt, 1954) where the average concentration of platinum ranges between 0.001 and 0.005 mg/kg (Mason, 1966, Bowen, 1979).

Platinum is found both in its metallic form and in a number of minerals. The principal minerals are: sperrylite, $PtAs_2$; cooperite, $(Pt,Pd)S$; and braggite, $(Pt,Pd,Ni)S$. Primary deposits are associated with ultrabasic, rather than silicic, rock formations. Economically important sources exist in the Bushveld Igneous Rock Complex in Transvaal, Republic of South Africa, and in the Noril'sk region of Siberia, the Kola Peninsula, and in the Nishnij Tagil region of the Urals, USSR. The platinum content in these deposits is between 1 and 500 mg/kg. In the Sudbury district of Canada, platinum metal is contained in copper-nickel sulfide ores at an average concentration of 0.3 mg/kg but is concentrated to more than 50 mg/kg during the refining of copper and nickel. In the USA, there is a platinum-palladium mine in the Stillwater Complex area, Montana (NAS, 1977; Renner, 1979).

Small amounts of platinum are also mined from secondary or placer deposits in the USSR (Ural Mountains), Colombia, USA (Alaska), Ethiopia, and the Philippines. In these deposits platinum is present in the form of metallic alloys of varied composition (NAS, 1977).

3.2 Anthropogenic sources

3.2.1 Production levels and processes

3.2.1.1 World production figures

World mine production of platinum-group metals, 40-50% of which is platinum, has steadily increased during the last two decades. In 1971 production was 127 tonnes (51-64 tonnes platinum) and in 1972 it was 132 tonnes (53-66 tonnes platinum) (Butterman, 1975). In 1975, automobile exhaust gas catalysts were introduced in the USA in order to meet the stringent emission limits for carbon monoxide, hydrocarbons, and nitrogen oxides set by the Federal Clean Air Act. In Japan, the automobile catalyst was introduced at the same time. As a consequence, world production of PGM increased to 179 tonnes (72-90 tonnes platinum) in 1975, reaching a plateau of between 200 and 203 tonnes per year (80-102 tonnes platinum) during the period 1977-1983 (Loebenstein, 1982, 1988).

From 1984 onwards world production increased, apparently in response to the anticipated demand in Western Europe where automobiles are being increasingly fitted with catalytic converters. In 1987, world mine production of PGM amounted to about 270 tonnes (108-135 tonnes platinum) (Loebenstein, 1988).

The future demand for platinum depends on improvements in engine technology and emission control, but can be expected to increase further during the coming years. Data on platinum demand are presented in section 3.2.2.

3.2.1.2 *Manufacturing processes*

Most native placer platinum is recovered by dredging and, in less developed areas, by small hand operations. The copper and nickel sulfide ores are mined by large-scale underground methods and concentrated by flotation (Stokinger, 1981).

The isolation of pure platinum metal from raw materials involves two principal stages: (i) extraction of a concentrate of precious metals from the ore; (ii) refining the concentrate to separate the platinum-group metals from each other and purify them. These processes require sophisticated chemical technology and include precipitating crystallization and liquid-liquid extraction, often combined with redox reactions to change the oxidation state of the metals. Further processes involve halogenation and reduction reactions at annealing temperatures and special distillations (Renner, 1984).

Potential health hazards of exposure to soluble platinum salts are encountered during the later stages of the refining process. After dissolving platinum, palladium, and gold with aqua regia or Cl_2/HCl and the subsequent precipitation of gold by addition of ferrous salts, ammonium chloride is added to precipitate ammonium hexachloroplatinate, $(NH_4)_2[PtCl_6]$. After several purification processes there is a second precipitation of this complex salt, which is then filtered off, dried and finally calcined to yield a spongy mass of platinum metal having purity of 99.95-99.99%. This can be further purified by a cationic exchange technique (NAS, 1977; Stokinger, 1981).

Secondary sources in substantial quantities come from the reclamation of scrap and used equipment, particularly industrial catalysts. The recycling of platinum-group metals from automobile catalysts is also increasing (see section 4.3). In principle, the recycling of platinum involves the same wet-chemical and melting processes that are applied to its production from ores (Renner, 1984).

3.2.1.3 Emissions from stationary sources

a) Production

Data on emissions of platinum during production are not available.

b) Stationary catalysts

During the use of platinum-containing catalysts, platinum can escape into the environment in variable amounts, depending on the type of catalyst. Of the stationary catalysts used in industry, only those employed for ammonia oxidation emit major amounts.

The loss of platinum from ammonia oxidation gauzes during nitric acid production depends on the operating pressure. An average figure is 0.15 g/tonne of nitric acid (Sperner & Hohmann, 1976). Of this apparent loss, 70-85% is recovered on gold-palladium catchment gauzes, reducing the loss to 0.03 g/tonne (Anon., 1990a). The production of nitric acid in the USA in 1989 was 7 247 837 tonnes (Anon., 1990b). Thus the amount of platinum

Sources of Human and Environmental Exposure

"lost" in 1989 in the USA is calcu-lated to be 217 kg. This is the maximum amount that could be dissolved or suspended as a colloid in the nitric acid and, thus, could be introduced into the environment if the nitric acid is used in fertilizer production.

3.2.1.4 Emissions from automobile catalysts

Automobile catalysts are mobile sources of platinum. Although these catalysts are designed to function for 80 000 km or more (Koberstein, 1984), some loss of platinum can occur due to mechanical and thermal impact. The data on platinum emissions from automobile catalysts are very limited.

In the mid 1970s unrealistically high assumptions were made for platinum loss. Brubaker et al. (1975) estimated the loss to be about 12 μg Pt/km, which would mean a total loss of approximately 1 g after 80 000 km.

Experimental data show much lower emission rates. Malanchuk et al. (1974) found a platinum concentration of 0.029 μg/m^3 in an inhalation chamber that was fed by catalysed engine exhaust. On the basis of the chamber volume, flow rate, and the speed simulated on the engine test stand, an emission rate of 0.39 μg/km was calculated. In another US EPA study, Sigsby (1976) did not detect platinum in particulate exhaust emissions (< 5 μm) at a detection limit of 0.06 μg/g. In exhaust dilution tunnels, platinum was detected in larger particles in the range of 0.034 to 635 μg/g sample; whole or fragmented pellets contained the highest concentrations.

Reliable emission data for the pellet-type catalyst come from a study conducted by the General Motors Corporation (Hill & Mayer, 1977), in which emission rates as well as the soluble fraction were determined by a radiometric method. Platinum emission was found to be 0.8 to 1.2 μg per km travelled in low-speed runs (starts and stops, maximum speed 48 km/h) and 1.9 μg per kilometre travelled in high-speed runs (96 km/h). It should be noted that these results relate to the first 250 km of catalyst life. Lower loss rates would be expected with increasing age of the catalyst. Of the particles collected, 80% had particle diameters greater than 125 μm. Experiments with an engine test stand using laboratory prepared catalysts indicated that about 10% of the platinum emitted is water soluble. However, the statistical significance of these results was not reported. Even so, these emission data provide the best basis for the estimation of expected ambient air concentrations resulting from the introduction of pellet catalysts (see section 5.1.1). However, this type of automobile catalyst is no longer used on new cars in the USA, and has never been used in Europe where only monolithic catalysts are on the market.

Emission data are available concerning the new generation monolith-type catalyst. In Germany the Fraunhofer Institute of Toxicology and Aerosol Research (König et al., 1989, König & Hertel, 1990) has conducted engine test stand experiments as part of a programme of the Ministry of Research and Technology for assessing the relative risk of this new man-made environmental source (GSF, 1990). First results indicated that platinum emission is lower by a factor of 100 than in the case of pelleted catalysts: at a simulated speed of 100 km/h, total loss from a three-way catalyst was measured, using the AAS method, to be on average about 17 ng/m^3 in the exhaust gas (König et al., 1989). In further experiments this value was validated (König & Hertel, 1990): the mean platinum emission from two catalysts was found to be 12 and 8 ng/m^3, respectively. As shown in Table 5, platinum emission seems to be temperature dependent. At an exhaust gas temperature of 690° C and a simulated speed of 140 km/h, about 35-39 ng/m^3 was found in the exhaust gas. The mean aerodynamic diameter of the particles collected after the muffler (silencer) on a Berner impactor varied between 4 and 9 μm. Preliminary results indicated that approximately 10% of the total platinum penetrated a depth-type filter to be trapped in the condensate (König et al., 1989), but this single measurement could not be confirmed by subsequent determinations where the platinum content in the condensate was below the detection limit (0.1 ng/ml) (König & Hertel, 1990).

Schlögl et al. (1987) analysed microparticles emitted from automobile exhaust and collected on several conducting surfaces. In experiments with diesel and gasoline engines equipped with catalysts, they found detectable traces of platinum. In diesel engine exhaust it was pre-

Sources of Human and Environmental Exposure

Table 5. Mean platinum emissions from two monolith catalysts (1 and 2) at different engine test stand runs[a]

Simulated speed (km/h)	Number of samples	Exhaust gas temperature (° C)	Platinum emission					
			Exhaust gas (ng/m³)		ng per km travelled[b]		Mean aerodynamic diameter (μm)	
			(1)	(2)	(1)	(2)	(1)	(2)
60	18	480	3	4	2	3	6	9
100	39	600	12	8	10	8	4	6
140	18	690	39	35	39	35	6	8

[a] Adapted from König et al. (in press)
[b] Calculated assuming that on average 10 m³ exhaust gas is emitted per litre gasoline and a gasoline kilometrage of 7, 8, and 10 litres per 100 km travelled, respectively.

sumed that most platinum would be in the oxidation state 0 (platinum black). A small part was found to be Pt(IV), probably in the oxide form. The platinum emission from gasoline engines showed a photoemission spectrum indicating that platinum is probably emitted mostly in the form of surface oxidized particles.

3.2.2 Uses

The principal use of platinum derives from its special catalytic properties. Further applications in industry are related to other outstanding properties, particularly resistance to chemical corrosion over a wide temperature range, high melting point, high mechanical strength, and good ductility. Platinum has long been known to have excellent catalytic properties. Before the introduction of catalytic converters in automobiles, most of the platinum was used as a catalyst in hydrogenation, dehydrogenation, isomerization, cyclization, dehydration, dehalogenation, and oxidation reactions. One of its major industrial uses is for naphtha-reforming to upgrade catalytically the octane rating of gasoline. Other catalytic uses are in ammonia oxidation to produce nitric acid, hydrogen cyanide manufacture, the reduction of nitro groups and, in the automobile catalyst application, the conversion of carbon monoxide to carbon dioxide and nitrogen oxide to nitrogen and water (NAS, 1977; Stokinger, 1981).

As shown in Table 6, in the USA in 1973, before the introduction of the automobile catalyst, most of the platinum was used for catalytic purposes in the chemical and petroleum industry. In 1987 the use pattern had completely changed and 71% of the platinum sold was used by the automobile industry. In 1987, a typical USA car catalyst contained about 1.77 g of platinum and 10.6 million vehicles with catalysts were produced (Loebenstein, 1988); this accounts for the 18.8 tonnes shown in Table 6.

Table 6. Platinum sales to various types of industry in the USA before and after the introduction of automotive catalytic converters[a]

Industry	1973		1987	
	kg/year	% of total	kg/year	% of total
Automobile	-	-	18 817	71.3
Chemical	7434	36.3	1920	7.5
Petroleum	3844	18.8	739	2.8
Dental and medical	868	4.2	479	1.9
Electrical	3642	17.9	1821	7.1
Glass	2255	11.0	285	1.1
Jewellery and decorative	697	3.4	177	0.7
Miscellaneous	1732	8.5	1430	5.6
Total	20 472	100	25 668	100

[a] From: Butterman (1975); Loebenstein (1988)

Tables 7 and 8 show the platinum demand by application in the Western world, also reflecting the increased demand during recent years. In 1989, total demand was 90 tonnes.

Platinum oxidation catalyst technology, developed to reduce automobile exhaust emissions, has been extended to other environmental control applications such as the reduction of carbon monoxide and hydrocarbon emissions from large gas turbines (Jung & Becker, 1987) and the

Table 7. Western-world platinum demand (kg/year) by application[a]

	1980	1981	1982	1983	1984	1985	1986	1987	1988	1989
Automobile catalyst										
gross	19 278	18 144	18 569	18 285	23 814	27 783	32 318	35 579	37 563	41 107
recovery	0	0	283	850	1276	1984	2551	3260	4536	4961
Chemical	7371	7087	7371	6946	7371	6379	5528	5528	4536	4536
Electrical	5953	5245	4819	4961	5386	5670	5103	5103	5245	5528
Glass	3969	2835	2410	2977	3969	3969	2551	3402	3685	3969
Investment										
small	0	0	1276	2551	4819	7371	12 757	6095	9355	3685
large	4536	5528	3260	1843	4252	4819	3544	7796	8505	850
Jewellery	15 876	21 404	21 687	20 270	21 971	22 963	24 097	28 066	33 452	36 996
Petroleum	3685	3969	1843	567	425	425	567	1559	1417	2126
Other	5386	4678	4819	4252	3827	2835	3685	3402	3402	3260
Total	66 054	68 889	65 771	61 802	74 559	80 230	80 511	93 270	102 624	97 096

[a] From Johnson Matthey (1990)

Table 8. Regional platinum demand (kg/year) by application[a]

	1980	1981	1982	1983	1984	1985	1986	1987	1988	1989
Japan										
Automobile catalyst										
gross[b]	5953	5386	4819	4819	4819	5953	7229	8788	9355	10 064
recovery[c]	0	0	0	0	0	0	142	425	709	709
Chemical	283	283	283	283	425	425	425	425	425	425
Electrical	425	425	567	567	50	1134	1276	1276	1276	1417
Glass	1134	1417	1276	1701	2126	1701	850	1276	1276	1134
Investment										
small	0	0	0	142	425	992	992	1701	3260	992
large	4536	5528	3260	1843	4252	4819	3544	7796	8505	850
Jewellery	12 474	17 718	17 577	15 876	17 718	19 136	20 979	25 515	30 050	32 602
Petroleum	425	425	425	425	567	425	0	0	0	0
Other	1417	1417	1559	1276	1134	850	567	425	425	425
Total	**26 647**	**32 599**	**29 766**	**26 932**	**32 316**	**35 435**	**28 632**	**46 777**	**53 863**	**47 200**

49

Table 8 (contd).

	1980	1981	1982	1983	1984	1985	1986	1987	1988	1989
North America										
Automobile catalyst										
gross	12 474	12 190	12 899	12 757	18 002	19 845	21 120	19 561	19 561	20 412
recovery	0	0	0	850	1276	1984	2410	2835	3827	4252
Chemical	3260	1417	2268	2835	2835	2126	1843	1559	1559	1559
Electrical	4111	1984	1984	2551	2693	2268	1843	1843	1843	2126
Glass	1417	567	283	425	850	1134	709	709	709	850
Investment	0	0	1134	1134	850	3685	8505	2410	2410	1559
Jewellery	425	425	425	425	425	425	425	425	425	567
Petroleum	3969	1559	567	425	425	283	283	425	425	1134
Other	2126	1701	567	709	992	850	1417	1417	1417	1417
Total	27 782	19 843	20 127	20 411	25 796	28 632	33 735	25 514	24 522	25 372

Table 8 (contd).

	1980	1981	1982	1983	1984	1985	1986	1987	1988	1989
Rest of Western world, including Europe										
Automobile catalyst										
gross	850	567	567	709	992	1984	3969	7229	8647	10 631
recovery	0	0	0	0	0	0	0	0	0	0
Chemical	3827	5386	4819	3827	4111	3827	3260	3544	2551	2551
Electrical	1417	2835	2268	1843	1843	2268	1984	1984	1984	1984
Glass	1417	850	850	850	992	1134	992	1417	1701	1984
Investment	0	0	142	1276	3544	2693	3260	1984	3685	1134
Jewellery	2977	3260	3685	3969	3827	3402	2693	2126	2977	3827
Petroleum	709	1984	850	283	567	283	283	1134	992	992
Other	1843	1559	2693	2268	1701	1134	1701	1559	1559	1417
Total	**11 622**	**16 441**	**15 874**	**14 459**	**16 443**	**16 159**	**18 142**	**20 977**	**24 096**	**24 520**

a From: Johnson Matthey (1990)
b Gross automobile catalyst demand is purchase of platinum by the auto industry for the manufacture of automobile catalysts.
c Automobile catalyst recovery is platinum recovered from catalytic converters removed from scrapped automobiles.

transformation of hydrogen molecules into active hydrogen atoms to reduce chlorohydrocarbons such as trichloroethylene to ethane in water (Wang & Tan, 1987).

Platinum and platinum-rhodium alloys have many high-temperature uses. Thermo-electrical applications arise from the simple and stable relationship between resistance and temperature that platinum exhibits over a wide temperature range. This explains its use in platinum resistance thermometers, thermocouples, and strain gauges. The high melting point of platinum and its resistance to oxidation and many chemicals has led to its use in vessels in the glass-making industry and in the fabrication of spinning jets and bushings for the production of viscose rayon and fibreglass, respectively. It is also used for laboratory ware, such as crucibles, combustion boats, and the tips of tongs. Ships' hulls, propellers, and rudders are protected against corrosion by "cathodic protection" using platinum-clad anodes (NAS, 1977).

Platinum and/or its alloys have been used in electric contacts for relays and switchgears for a variety of reasons, including hardness and good conductivity. Many printed circuits are made using preparations that contain platinum. Electrochemical platinum electrodes have been used in preparative chemistry, since they support many oxidative reactions although they resist oxidation themselves (NAS, 1977).

A major use of platinum is in jewellery for making rings and settings. Platinum is also used to produce a silvery lustre on ceramic glazes (NAS, 1977).

In dentistry, platinum is used in gold-platinum-palladium alloys to raise the melting-point range and increase the strength. However, this use is decreasing, since platinum is being replaced by other materials including palladium (Anusavice, 1985; NAS, 1977).

Platinum has an important role in neurological prostheses, i.e. surgically implanted microelectronic devices, such as implants for treating incontinence, or for recovering some use of paralysed limbs following spinal accidents (Donaldson, 1987).

Platinum-iridium electrodes are used for long-term electrode implantation for recording electrical activity and for stimulation in human tissues and organs, e.g., pacemakers (Theopold et al., 1981).

All these applications use platinum as a pure metal or in the form of alloys, but soluble platinum salts are also used in the manufacture of these products; e.g., hexachloroplatinic acid may be used in platinizing alumina or charcoal in catalyst production. A number of salts can be used in the electrodeposition of platinum, e.g., sodium hexahydroxyplatinate(IV), $Na_2[Pt(OH)_6]\cdot 2H_2O$, diamminedinitroplatinum(II), $[Pt(NO_2)_2(NH_3)_2]$, hydrogen dinitrosulfatoplatinate(II), $H_2[Pt(NO_2)_2SO_4]$, and tetraammineplatinum(II) compounds such as the hydrogenphosphate, sulfamate, citrate, and tartrate (Baumgärtner & Raub, 1988; Skinner, 1989).

Complexes of platinum, particularly *cis*-diamminedichloroplatinum(II) (cisplatin) (see footnote in section 1.2), have been used to treat cancer. In patients with testicular cancers, remissions rates of more than 90% have been achieved (Lippert & Beck, 1983).

4. ENVIRONMENTAL TRANSPORT, DISTRIBUTION, AND TRANSFORMATION

4.1 Transport and distribution between media

By comparison with other elements, platinum-group metals are distributed sparsely in the environment. Since platinum is so valuable, great care is taken to avoid significant loss during mining and refining processes, and during use and disposal of used platinum-containing objects. Up to 1984, about 1050 tonnes of platinum had been refined. Most of this has been used in the form of the metal and platinum oxides, which are practically insoluble in water, resistant to most chemical reactions in the biosphere, and do not volatilize into air (Renner, 1984).

Part of the platinum released into the air from automobile emissions (section 5) is deposited close to the roads and could be washed off by rain into rivers and coastal marine waters (Hodge & Stallard, 1986). However, only small amounts of platinum have been detected in environmental samples (see sections 5.1.2. and 5.1.3.).

Large amounts of metals including platinum can be transported in rivers draining major industrialized regions, leading to elevated platinum concentrations in sediments (section 5.1.3).

Platinum forms soluble complexes with ammonia, cyanide, amines, olefins, organic sulfides, and tertiary arsines. However, the level of these ligands in natural waters is insufficient to make platinum mobile (Fuchs & Rose, 1974).

Organic matter has a role as a vehicle for the transport of platinum and for bringing about its precipitation or concentration. There is a good correlation between high contents of platinum and organic carbon in polluted stream sediments of the Ginsheimer-Altrhine river, near Mainz, Germany (see section 5.1.2), and it is assumed that organic matter such as humic and fulvic acids binds platinum, aided perhaps by appropriate pH and redox potential conditions in the aquatic environment (Dissanayake, 1983).

Detailed information about the geochemical behaviour of platinum-group metals is available from the platinum mining area of Stillwater, Montana, USA (Fuchs & Rose, 1974). The mobility of platinum depends on pH, the redox potential, chloride concentrations in soil water, and the mode of occurrence of platinum in the primary rock. The relation between redox potentials and pH conditions indicates that platinum behaviour also depends on the kind of ore it is associated with. If bound in chromite, it has essentially no mobility in weathering because of the resistant character of chromite. On the other hand, platinum in the form of trace mineral inclusions in sulfides is readily released by oxidation during weathering. Calculated relations between pH and redox potential indicate that increased chloride concentrations in soil water will promote mobility. Thus, platinum will be mobile only in extremely acid waters or those with a high chloride level (Fuchs & Rose, 1974).

In twigs from four limber pines (*Pinus flexilis*) in the platinum mining area of Stillwater, the platinum concentrations were the same as in the adjacent soil. It was concluded that limber pine does not concentrate platinum, probably due to the limited mobility of platinum (Fuchs & Rose, 1974). However, high concentrations of platinum were found in the roots of nine horticultural crops (cauliflower, radish, snapbean, sweet corn, pea, tomato, bell pepper, broccoli, and turnip) grown in Hoagland's hydroponic culture solution containing platinum tetrachloride concentrations of 0.057, 0.57, or 5.7 mg/litre (Pallas & Jones, 1978; see section 7.3). For example, at the highest concentration, cauliflower and tomato roots contained 1425 and 1710 mg Pt/kg, respectively. Only pepper, cauliflower, and radish accumulated platinum in their tops, but to a very limited extent. From the data of Pallas & Jones (1978) it is not clear whether they differentiated between contamination of the root surface and true uptake of platinum. However, these results indicate that platinum can enter food crops but the bioavailability essentially depends on the solubility of the platinum species. It should be noted that the salt ($PtCl_4$) used by Pallas & Jones (1978) is soluble in water.

In the context of a German government programme (see section 3.2.1.4), Rosner et al. (1991) conducted engine test stand experiments with a three-way-catalyst-equipped engine (monolith-type catalyst) to determine platinum uptake by plants. Grass cultures (*Lolium multiflorum*) were placed in continuously stirred tank reactors and exposed to slightly diluted (1:10/20) exhaust gas for 4 weeks (8 h/day, 5 days/week). Using atomic absorption spectrometry for the measurement of platinum emissions (see section 2.4.3.8, König & Hertel, 1990), no platinum could be detected in the shoots at a detection limit of 2 ng/g dry weight.

4.2 Biotransformation

By analogy, platinum compounds may undergo biotransformation comparable to processes described for other metals. The biomethylation of platinum compounds, i.e. $[Pt(IV)Cl_6]^{2-}$, $[Pt(IV)(CN)_4Cl_2]^{2-}$, $[Pt(IV)(CN)_5Cl]^{2-}$, and $[Pt(IV)(SO_4)_2]$, has been established only in *in vitro* test systems (Taylor, 1976; Wood et al., 1978; Fanchiang et al., 1979; Taylor et al., 1979; Fanchiang, 1985).

Methylcobalamin (MeB_{12}) reacts with Pt(II) and Pt(IV) complexes to give a methylated platinum compound. Agnes et al. (1971) reported that this reaction requires the presence of platinum in both oxidation states. Spectrophotometric measurements showed the consumption of one mole of $[Pt(IV)Cl_6]^{2-}$ per mole of MeB_{12}, $[Pt(II)Cl_4]^{2-}$ being required only in catalytic quantities. Aquocobalamin (aquo-B_{12}) and methylplatinum were shown to be the products of the reaction (Taylor & Hanna, 1977).

From these laboratory data produced under abiotic conditions it is not, however, possible to conclude that microorganisms in the environment are able to biomethylate platinum complexes.

4.3 Ultimate fate following use

The value of platinum-group metals has greatly increased and methods for their recovery from spent catalysts are of economic importance.

Platinum metal has been successfully recycled from used chemical and petroleum catalysts for many years, but many companies are still trying to find a successful formula for retrieving it from automobile catalysts. The latter accounts for more than 30% of the total platinum-group metal consumption in the USA. The US Office of Technology calculated that if 50-60% of catalytic converters were recovered for their metal value, about 7717 kg platinum per year could be reclaimed in 1990. However, currently only between 25 to 40% of the used converters are being reclaimed (Agoos, 1986). According to another estimate, 5443 kg of platinum was recovered in 1989 from automobile catalysts, of which 4666 kg was recovered in the USA (Johnson Matthey, 1990).

In contrast to automobile catalysts, almost 100% of spent reforming and gauze catalysts are collected for their metal value. This is based on their much higher platinum metal content (Agoos, 1986).

5. ENVIRONMENTAL LEVELS AND HUMAN EXPOSURE

5.1 Environmental levels

5.1.1 Ambient air

Few measurements of platinum ambient air concentrations have been reported. Results obtained before the introduction of cars with catalytic converters can serve as a baseline. Air samples taken near freeways in California, USA, and analysed using atomic absorption spectrometry were below the detection limit of 0.05 pg/m^3 (Johnson et al., 1975; 1976).

No platinum could be detected in two air samples collected by Ito & Kidani (1982) in an industrial area of Nagoya, Japan, in 1981.

Close to city roads in Frankfurt, Langenbrügge, Germany, the platinum air concentrations (particulate samples) were measured in 1989 to be between $\leq$ 1 and 13 pg/m^3. In rural areas the concentrations were $\leq$ 0.6-1.8 pg/m^3 (Tölg & Alt, 1990). At the time of these measurements, few German cars were equipped with catalysts. Thus, these levels virtually reflect background levels.

Rosner & Hertel (1986) estimated ambient air concentrations for different scenarios, based on dispersion models used by US EPA (Ingalls & Garbe, 1982) and on the emission data of Hill & Mayer (1977) (see section 3.2.1.4). As shown in Table 9, total platinum concentrations near and on roads could range from 0.005 to 9 ng/m^3. Estimates for parking and personal garages were also made, based on an assumed emission rate of 1 µg/min for total platinum, but this is definitely an overestimate. It can be assumed that the emission of platinum depends on the exhaust gas temperature. At idling or very low speed conditions, emissions are expected to be negligible (see section 3.2.1.4).

As described in section 3.2.1.4, emission data indicate that the total platinum emission of a monolith-type catalyst is probably lower by a factor of 100 than that of a pellet-type catalyst. Assuming an average emission rate of approximately 20 ng/km (see section 3.2.1.4) and applying the same dispersion models, the theoretical ambient air concentrations would be lowered to the picogram to femtogram per m^3 range (see Table 9).

Table 9. Estimated ambient air concentrations of total platinum at various exposure conditions, based on an emission rate of 2 μg/km from the pelleted catalyst and 0.02 μg/km from the monolithic three-way catalyst

Exposure situation[a]	Ambient Pt concentration (ng/m^3)	
	Pelleted catalyst	Monolithic catalyst
Roadway tunnel		
Typical	4	0.04
Severe	9	0.09
Street canyon (sidewalk receptor)		
Typical a) 800 vehicles per h	0.1	0.001
Typical b) 1600 vehicles per h	0.3	0.003
Severe a) 1200 vehicles per h	0.5	0.005
Severe b) 2400 vehicles per h	0.9	0.009
On expressway		
Typical	0.7	0.007
Severe	1.6	0.016
Beside expressway (short-term)		
Severe 1 m	1.3	0.013
10 m	1.1	0.011
100 m	0.3	0.003
1000 m	0.04	0.0004
Beside expressway (annual)		
Severe 1 m	0.2	0.002
10 m	0.15	0.0015
100 m	0.04	0.0004
1000 m	0.005	0.00005

[a] Calculations based on dispersion models used by US EPA; "Typical/severe" depends on wind conditions and road width (Ingalls & Garbe, 1982)

Hodge & Stallard (1986) analysed roadside dust deposited in San Diego, California, USA. At the edge of a major freeway (154 000 vehicles/day), dust samples contained the highest concentration (680 μg Pt/kg dry weight; 680 ppb). At a distance of about 34 m, the platinum content of 100 μg/kg was about 7 times lower. At the edge of another heavily used freeway (96 000 vehicles/day) platinum content was 250 µg/kg, while with less heavy traffic (14 000 vehicles/day) 260 and 300 µg/kg were found in two dust samples. The lowest concentrations, 37 and 60 µg/kg, were found in samples collected from plants growing in the yards of houses located on highly used road. The platinum concentration was not correlated with the lead concentration. However, the samples with the highest platinum concentrations also had the highest lead values. Although the number of samples was limited, the results indicate that automobile catalysts release platinum. However, it should be noted that platinum emissions from pelleted catalysts were probably responsible for the concentrations reported and that the use of monolith catalysts should result in much lower platinum concentrations in the roadside environment.

5.1.2 Water and sediments

In a study to determine baseline levels of platinum, Johnson et al. (1976) analysed tap water samples collected in Lancaster and Los Angeles, California, USA. No platinum was found at a detection limit of 0.08 µg/litre. In tap water (probably only one sample) from Liverpool, United Kingdom, a platinum content of 0.06 µg/litre was determined by adsorptive cathodic stripping voltametry (Van den Berg & Jacinto, 1988).

Investigations of platinum concentrations in Lake Michigan sediments led to the conclusion that platinum has been deposited over the past 50 years at a constant rate. Concentrations at sediment depths of 1-20 cm varied between 0.3 and 0.43 µg/kg dry weight (Goldberg et al., 1981). In comparison, lead concentrations have markedly increased in the sediment due to increased emissions from industry and motor traffic.

Lee (1983) noted a rapid increase in the palladium contents of the sediments from the Palace Moat, Tokyo, Japan, between 1948 and 1973 and attributed it to the introduction of car catalysts. However, this is not conclusive as the palladium content in the sediment had already begun to increase in 1964-1965, before the introduction of the catalytic converter, and even in 1973 only a few cars were equipped with converters.

Dissanayake et al. (1984) determined platinum concentrations in the sediments of a cut-off channel of the Rhine river near Mainz, Germany. Sediment samples from this highly polluted river were sieved and the < 2 µm fraction was analysed by flameless AAS. The platinum concentrations in 12 samples collected at different sites varied over a wide range. In four samples no platinum was detected, while eight samples contained between 730 and 31 220 µg/kg (dry weight). This is higher by a factor of up to 15 000 compared to unpolluted average North Sea sediments. The high variation was attributed to differences in pH and redox conditions. The extremely high concentrations appeared at the interface between an extremely reducing and an oxidizing aquatic environment that provided, together with a pH of 6.6-7.8, optimum conditions for the formation of metal-organic complexes. The sample containing 31 220 µg Pt/kg also contained the highest concentration of palladium (4000 µg/kg). The gold content (100-400 µg/kg) had a relatively uniform distribution, but was also indicative of a high state of pollution.

Using a more sensitive graphite furnace AAS method, Goldberg et al. (1986) detected very low platinum concentrations in sea water. Samples of filtered water (0.45-µm filter) from the open Eastern Pacific Ocean showed an increase in platinum concentration with depth from surface values of around 100 to a value of 250 pg/litre at 4500 m. Similar concentration profiles were obtained in unfiltered sea water taken from the California Borderline region (Hodge et al., 1985). Sea-water samples analysed by Van den Berg & Jacinto (1988) were also within this concentration range. A deep-sea and a shallow-water sample from the Indian Ocean contained 154 and 37 pg/litre, respectively, whereas sea water of coastal origin contained 332 pg/litre. It should be noted that these were only single samples.

In sediment cores from the Eastern Pacific taken to a depth of 6-22 cm in carbonate and siliceous ooze, platinum concentrations varied between 1.1 and 3 µg/kg (dry weight basis). Lower concentrations (0.3 µg/kg) were reported in the Santa Barbara Basin (Hodge et al., 1985). The highest concentration (21.9 µg/kg) was found in pelagic ocean sediments (Hodge et al., 1986).

In several investigations, the platinum content of seamount ferromanganese nodules or crusts was studied. In deep-sea nodules from the Northwest Pacific nodule belt, platinum concentrations from ≤ 5 to 145 µg/kg were found (Agiorgitis & Gundlach, 1978).

Platinum values in ferromanganese seamount crusts from the Central Pacific were much higher and varied between 140 µg/kg at 3780 m and a maximum of 880 µg/kg at a depth of 1120 m (Halbach et al., 1984). Both platinum and nickel concentrations correlated positively with manganese content and led to the conclusion that platinum and nickel are incorporated in the manganese oxide fraction. It was suggested that the high platinum concentration in the crusts is derived directly from sea water by a process of specific adsorption onto colloidal particles of hydrous manganese oxide, which has a negative surface charge in sea water.

In a further investigation, platinum concentrations in ferromanganese minerals from various localities were found to vary between 6 and 940 µg/kg (Goldberg et al., 1986). In manganese nodules obtained at depths of between 1700 and 4200 m in the Pacific Ocean, platinum concentrations varied between 138 and 940 µg/kg (Hodge et al., 1986).

5.1.3 Soil

Few measurements of platinum in soil have been reported. In the baseline study of Johnson et al. (1976), all surface soil samples collected near freeways in California, USA, and in a mining area in Sudbury, Canada, were below the detection limit of 0.8 µg/kg.

In the USA, the National Academy of Sciences (NAS, 1977) estimated the accumulation of platinum in roadside environments on the basis of an emission rate of 1.9 µg per km from cars equipped with catalytic converters and a frequency of 5000 cars per day. Assuming that all emitted platinum was localized near the freeway in the topsoil (uniformly distributed about 30 cm deep over a width of about 90 m and a length of 1.6 km, with a soil density of 1.5 g/cm^3), a platinum concentration after 10 years of 8 µg/kg could be expected.

5.1.4 Food

Hamilton & Minski (1972/1973) estimated a total daily platinum intake of less than 1 µg/day, based on an analysis of a United Kingdom total-diet sample and 1963 United Kingdom consumption and population figures. No data were given on the platinum content of the foods analysed.

5.1.5 Terrestrial and aquatic organisms

Fuchs & Rose (1974) analysed samples of twigs from four limber pines (*Pinus flexilis*) in the Stillwater mining area, Montana, USA. Three samples contained between 12 and 56 µg Pt/kg (ash weight), while one contained platinum at a level below the detection limit. The content of the adjacent soils was also in this range, so that no evidence for accumulation could be derived from these limited data (see also section 4.1).

Using neutron activation analysis (section 2.4.3.2) Valente et al. (1982) measured the following platinum concentrations in isolated samples of plants from an ultrabasic soil: *Fragaria virginiana*, 830 µg/kg (dry weight); *Prunella vulgaris*, 440 µg/kg; *Aspidotis densa*, 100 µg/kg.

In marine macroalgae the following platinum concentrations (on a dry weight basis) were found near La Jolla, California, USA (Hodge et al., 1986): red algae *Prionites australis* and *Opuntiella californica*, 0.19 and 0.08 µg/kg, respectively; brown algae *Macrocystis pyrifera* and *Pterygophora californica*, 0.22 and 0.32 µg/kg, respectively.

5.2 General population exposure

Two studies were conducted in the USA to establish baseline levels of platinum in the tissues and body fluids of the general population prior to the introduction of automobile catalysts.

Johnson et al. (1975, 1976) analysed autopsy tissue samples from 10 people, 12 to 75 years old, who died from a variety of causes in Southern California. All samples taken from liver, kidney, spleen, lung, muscle, and fat were below the detection limits (0.2-2.6 µg/kg wet weight). Samples collected from 282 people from Southern California living near a heavily used urban freeway (Los Angeles) or in a desert area near Lancaster also showed platinum concentrations below the detection limits (blood, < 31 µg/litre; urine, < 0.6 µg/litre; hair, < 50 µg per kg; faeces, < 2 µg/kg). Only in pooled blood samples were detectable concentrations measured, i.e. 0.49 µg/litre in the Los Angeles group and 1.8 µg/litre in the Lancaster group.

In a second study, tissue samples were taken from autopsied individuals from Southern California (95 people) and New York (2 people), who had not been knowingly exposed to platinum either occupationally or by medical treatment (Duffield et al., 1976). In 42 individuals no platinum was detected. Of the 1313 samples collected, only 62, i.e. 5%, had detectable concentrations of platinum ranging from 0.003 to 1.46 mg/kg wet weight (mean 0.16 mg/kg, median 0.067 mg/kg). Table 10 shows the frequency of platinum detection in the various tissue samples. The frequency of occurrence was taken as a measure of the distribution of platinum among various body organs. Platinum was frequently found in subcutaneous fat. This is surprising, as most platinum compounds are regarded as lipid-insoluble. Other target sites were kidney, pancreas, and liver. However, the analytical accuracy has been questioned and contamination of the samples suspected (NAS, 1977), because the baseline levels found by Johnson et al. (1976) were at least one order of magnitude lower. The problem of questionable analytical reliability reflects the difficulties in interpreting data on trace levels of platinum in the environment and in human tissues and body fluids.

New data have been provided by Nygren et al. (1990). Using absorptive voltametry (see section 2.4.3.5), the background levels of platinum in human blood were found to be in the range of 0.1-2.8 µg/litre (median 0.6 µg per litre). These results were verified by inductively coupled plasma mass spectrometry using gold as an internal standard.

Table 10. Distribution of tissue samples with detectable platinum[a]

	Number of samples analysed	Samples with detectable platinum	
		No.	%
Subcutaneous fat	74	10	14
Kidney	91	11	12
Pancreas	84	10	12
Liver	90	10	11
Brain	9	1	11
Gonad	53	5	9
Adrenal	60	3	5
Muscle (psoas)	97	4	4
Aorta (descending)	92	3	3
Heart (left ventricle)	82	2	2
Spleen	52	1	2
Prostate/uterus	63	1	2
Thyroid	73	1	1
Lung	95	0	0
Vertebra (lumbar)	94	0	0
Rib (fifth)	97	0	0
Femur	57	0	0
Clavicle	30	0	0
Hair, scalp	9	0	0
Hair, pubic	1	0	0
	1303	62	5

[a] From: Duffield et al. (1976)

5.3 Occupational exposure during manufacture, formulation, or use

Occupational exposure occurs during the mining and processing of platinum. However, the most common current occupational exposure to soluble platinum compounds is through inhalation in platinum refining and catalyst manufacture.

Many countries have set occupational exposure limits. For example, in the USA, the time-weighted Threshold Limit Value (TWA-TLV) for daily occupational exposure has been established for soluble platinum salts at 2 μg Pt/m^3 (ACGIH, 1980, 1990). Many countries have adopted this ACGIH value. In addition ACGIH (1980, 1990) recommended a Threshold Limit Value of 1 mg/m^3 for platinum metal. In the United Kingdom an occupational exposure limit (8-h TWA) of 5 mg/m^3 has been proposed for platinum metal as total inhalable dust (Health and Safety Executive, 1990).

The published data base for platinum concentrations at the workplace is meagre. Due to analytical shortcomings older data are not considered reliable. In an early investigation (Fothergill et al., 1945), a platinum content of less than 5 µg/m³ in the atmosphere in the immediate neighbourhood of a refinery was measured using particle filters. In the dry salts handling area, platinum concentrations as high as 70 µg/m³ were found. In another investigation (Hunter et al., 1945), the platinum content in the atmosphere at various points in four refineries was estimated. At most points concentrations varied between 1.6 and 5 µg/m³. Higher concentrations were found in the neutralization of platinum salts (20 µg/m³), sieving spongy platinum (400-900 µg/m³), and crushing ammonium chloroplatinate (1700 µg/m³).

Workplace measurements in a catalyst production plant in the USSR were reported to exceed an air concentration of 2 µg/m³ in 33% of the measurements (Gladkova et al., 1974).

In a cross-sectional survey (section 9.2), Bolm-Audorff et al. (1988) reported workplace measurements at a platinum refinery in the Federal Republic of Germany. In 1986, concentrations of between 0.08 and 0.1 µg/m³ were measured in the filter press area, but in other working areas platinum salt exposure was generally below the detection limit of 0.05 µg/m³. No data were given on the number of samples.

The results obtained during a four-month period of measurements in a US platinum refinery showed that workplace concentrations exceeded the occupational limit of 2 µg/m³ between 50 and 75% of the time (Brooks et al., 1990.

In samples of blood, urine, faeces, and hair from employees at a Canadian mine near Sudbury, platinum concentrations were below the limits of detection (0.1 µg per litre or 0.1 µg/kg). Tissue samples from three out of nine autopsies had detectable platinum concentrations in fat (4.5 µg/kg), lung (3.7 µg/kg), or muscle (25.0 µg per kg) (Johnson et al., 1976). However, since the three detectable concentrations were in individuals who, like the other six, showed no platinum concentrations in liver, kidney and spleen, sample contamination was suggested

(NAS, 1977). It was concluded that people who work in mining areas probably do not incorporate significant amounts of platinum into their body.

Blood samples collected from 61 refinery workers in New Jersey contained no measurable platinum (less than 1.4 µg/litre) (Johnson et al., 1976). However, platinum levels in 10% of the urine samples were above the detection limit of 0.1 µg/litre, the maximum reported value being 2.6 µg/litre.

Using the method of LeRoy et al. (1977), platinum serum levels in 11 platinum refinery workers with positive skin tests were analysed. These studies found serum platinum levels ranging from 150 to 440 µg/litre (mean = 240 µg/litre), the quantification limit being 100 µg per litre (Biagini et al., 1985).

A special case of possible occupational exposure is the handling of cisplatin and its analogues by pharmacy and nursing staff and other hospital personnel. In a study with two pharmacists (one male and one female) and eight female nurses, platinum levels in urine (0.6-23.1 µg per litre) were at the limit of sensitivity of the AAS method used and did not significantly differ from the controls (2.6-15.0 µg/litre). By comparison, the urine of cisplatin-treated patients contained on average 7 mg/litre (Venitt et al. 1984).

6. KINETICS AND METABOLISM

Most toxicokinetic data on platinum, both for experimental animals and humans, have been derived from studies with platinum complexes.

Moore et al. (1975c) studied the whole body retention, lung clearance, distribution, and excretion of ^{191}Pt in outbred albino rats (Charles River CD-1 strain) after single nose-only inhalation exposure to different chemical forms of platinum for 48 min. Particle concentration in the nose-only exposure chambers was approximately 5.0 mg per m^3 with ^{191}PtCl$_4$, 5-7 mg/m^3 with ^{191}Pt(SO$_4$)$_2$, and 7-8 mg/m^3 with PtO$_2$ and ^{191}Pt metal. The aerodynamic diameter was given as 1.0 μm for ^{191}PtCl$_4$ and ^{191}Pt(SO$_4$)$_2$; both aerosols were generated by a nebulizer. The ^{191}PtO$_2$ and ^{191}Pt metal aerosols (aerodynamic diameter not given) were generated by passing Pt(SO$_4$)$_2$ or PtCl$_4$, respectively, through a furnace tube and decomposing them at 600 °C. Whole body counts, showed that most of the inhaled ^{191}Pt was rapidly cleared from the body, followed by a slower clearance phase during the remaining post-exposure period. The whole body retention of ^{191}Pt was approximately 41, 33, 31, and 20%, respectively, of the initial body burden 24 h after exposure to ^{191}PtCl$_4$, ^{191}Pt(SO$_4$)$_2$, ^{191}PtO$_2$, and ^{191}Pt metal. After ten days, the body burden was only about 1, 5, 8, and 6%, respectively. This shows that there was only a slight difference between the clearance rates for the various chemical forms, although the clearance of ^{191}PtCl$_4$ seemed to be the fastest. Clearance from the lungs also reflected the two-phase pharmacokinetics in the whole body, with a fast clearance phase in the first 24 h followed by a slow phase with a half-time of about 8 days.

Excretion data from the study by Moore et al. (1975c) indicate that most of the ^{191}Pt cleared from the lungs by mucociliary action was swallowed and excreted via the faeces (half-time 24 h). A small fraction of the ^{191}Pt was detected in the urine, indicating that little was absorbed by the lungs and the gastrointestinal tract. However, no quantitative data were given.

As shown in Table 11, the portals of entry, lung and trachea, contained most of the platinum, i.e. 93.5% and 3.9%, respectively, of the total radioactivity (48 618 counts/g) 1 day after exposure. Of the other tissues analysed, highest levels were found in the kidney and bone, suggesting some accumulation in these organs. The low percentages of 1.5% and 0.6%, respectively, on day 1, reflect only a low accumulation tendency; no information on the statistical significance of these figures was provided.

Table 11. Radioactive ^{191}Pt distribution in the rat following inhalation exposure to platinum metal (7-8 mg/m^3, 48 min)^a

| | Mean counts/g wet weight after exposure for | | | |
	1 day	2 days	4 days	8 days
Blood	61	43	30	12
Trachea	1909	2510	738	343
Lung	45 462	28 784	28 280	23 543
Liver	52	46	37	17
Kidney	750	1002	906	823
Bone	281	258	231	156
Brain	5	3	1	0
Muscle	22	10	28	0
Spleen	39	73	23	5
Heart	37	58	23	5

^{a} From: Moore et al. (1975c)

In a comparative study on the fate of ^{191}PtCl$_4$ (25 µCi per animal) in rats following different routes of exposure (Moore et al., 1975a,b) retention followed the classical pattern. The highest retention was found after intravenous administration, the next highest after intratracheal, and the lowest after oral administration. For comparison, retention after inhalation was lower than after intratracheal administration. However, the total dose was much higher with inhalation (7000 µCi) than with intratracheal (25 µCi) administration. Only a minute amount of ^{191}PtCl$_4$ given orally was absorbed. Most of it passed through the gastrointestinal tract and was excreted via the faeces. After 3 days less than 1% of the initial dose was detected in the whole body. Following intravenous administration, ^{191}Pt was excreted in almost equal quantities in both faeces and urine but elimination was slower than after oral dosing. After 3 days, whole body retention was about 65% and after 28 days it was still 14% of the initial dose. By comparison, following intratrachael administration about 22% and 8%, respectively, were retained by the body after these periods (Moore et al., 1975a,b).

In the same studies the tissue distribution of ^{191}Pt was determined. After the single oral dose, the kidney and liver contained the highest concentrations, while in the other organs there were no elevated levels. In contrast, after intravenous administration ^{191}Pt was found in all tissues (Table 12). The high concentration of ^{191}Pt found in the kidney shows that once platinum is absorbed most of it collects in the kidney and is excreted in the urine. The liver, spleen, and adrenal gland also contained higher platinum concentrations than the blood. The lower level in the brain suggests that platinum ions probably cross the blood-brain barrier only to a limited extent (Moore et al., 1975a,b).

This was confirmed by Lown et al. (1980) in male Swiss mice given single intragastric doses of $Pt(SO_4)_2$ (144 or 213 mg Pt/kg body weight). Platinum levels in the blood were several times higher than in the brain. Clearance from the whole body was slower than in the rat studies. This could be due to species-specific differences. In addition, the mice received much higher doses than the rats. Lown et al. (1980) noted an enhancing effect of the higher dose on absorption.

In a long-term study, Holbrook (1977) found evidence that a platinum-binding protein is induced. Male Sprague-Dawley rats received platinum salts *ad libitum* either in the drinking-water or in the dry feed. The sequential platinum contents in the tissues analysed are shown in Table 13. The data demonstrate that the oral administration of water-soluble platinum compounds, i.e. $PtCl_4$ and $Pt(SO_4)_2$, results in accumulation of platinum in some organs, primarily the kidney. After 4 weeks, the platinum content of the kidney was about 8-fold higher than that of the liver and spleen, and at least 16-fold higher than in the blood and testis (except for the highest dose of $PtCl_4$). The total platinum intake after 4 weeks increased by 4.3 times and the platinum content in

Table 12. Radioactive ^{191}Pt distribution (counts/g wet weight) in the rat following a single intravenous dose of PtCl$_4$ (25 µCi/animal)[a]

Tissue	1 day		2 days		7 days		14 days	
	%	counts/g	%	counts/g	%	counts/g	%	counts/g
Blood	0.91	22 147	0.81	19 732	0.52	12 774	0.32	7921
Heart	0.48	11 819	0.50	12 201	0.36	8 805	0.19	4593
Lung	0.75	18 432	0.66	16 139	0.46	11 180	0.24	5770
Liver	1.51	36 848	1.28	31 274	1.05	25 732	0.19	4733
Kidney	6.65	162 227	6.59	160 656	5.66	138 010	1.24	30 195
Spleen	1.68	41 085	1.89	45 840	2.29	55 764	0.86	20 973
Pancreas	0.91	22 208	0.80	19 487	0.60	14 802	0.16	3973
Bone	0.53	13 146	0.52	12 800	0.37	8932	0.22	5440
Brain	0.05	1150	0.10	2485	0.02	595	0.01	265
Fat	0.18	4487	0.18	4501	0.13	3201	0.02	429
Testes	0.17	4186	0.27	6540	0.16	3873	0.06	1431
Adrenal	1.86	45 439	1.74	42 363	1.09	26 667	0.25	6190
Muscle	0.19	4798	0.19	4671	0.14	3441	0.09	2146
Duodenal segment	0.52	12 725	0.25	6044	0.16	4031	0.06	1410

[a] Adapted from: Moore et al. (1975a)

Table 13. Dietary levels, total platinum consumption, and platinum content of tissues after oral administration of platinum salts to rats[a]

Platinum salt	Pt consumption			Pt content (mg/kg wet weight; mean ± SE)[b]					
	Duration (weeks)	Dietary level (as Pt)	Total (mg Pt/rat)	Liver	Kidney	Spleen	Testis	Brain	Blood
PtCl$_4$	1	319[c]	59	2.2	4.8 ± 0.2	0.24			0.23
PtCl$_4$	4	319[c]	255	2.5 ± 0.9	33.7 ± 3.5	4.8 ± 1.5	1.5 ± 0.5	0.11 ± 0.07	2.1 ± 0.4
PtCl$_4$	4	1147[d]	743	3.2 ± 0.9	33.5 ± 6.3	3.1 ± 0.9	1.1 ± 0.4	< 0.02	1.5 ± 0.4
PtCl$_4$	4	2581[d]	1616	8.9 ± 1.2	32.4 ± 4.6	6.4 ± 3.0	1.7 ± 0.3	0.12 ± 0.08	1.6 ± 0.2
PtCl$_4$	13	106[c]	389	1.3 ± 0.3	14.9 ± 0.4	1.6 ± 0.3	0.94 ± 0.20	< 0.06	0.9 ± 0.08
Pt(SO$_4$)$_2$·4H$_2$O	1	106[c]	26	0.07	0.26 ± 0.02	< 0.02	< 0.04		0.05
Pt(SO$_4$)$_2$·4H$_2$O	1	319[c]	78	0.85	4.6	0.13		< 0.02	0.22
Pt(SO$_4$)$_2$·4H$_2$O	4	1147[d]	716	3.5 ± 0.4	43.4 ± 8.3	3.2 ± 0.5	1.1 ± 0.1	0.33 ± 0.18	1.6 ± 0.3
PtO$_2$	4	5808[d]	4308	< 2.2	< 2.2	< 0.02	< 0.07	< 0.02	< 0.04

[a] Adapted from: Holbrook (1977)
[b] Standard error (SE) is given for four values; only the mean is given when two values are available
[c] mg Pt/litre
[d] mg Pt/kg

kidney, spleen, and blood increased by at least 7 times as compared with the 1-week levels. It is notable that a more than 2-fold increase in the intake of platinum (after a 4-week consumption of $PtCl_4$ in the dry feed; 743 vs. 1616 mg Pt/rat) did not lead to an increase in the platinum content of the kidney, in contrast to the situation in the liver and spleen. This observation was not corroborated with $Pt(SO_4)_2$.

In contrast to the water-soluble salts, insoluble PtO_2 was only taken up in minute amounts even though the salt was administered in the diet at an extremely high level resulting in a total consumption of 4308 mg Pt/rat over the 4-week period (Holbrook, 1977).

Moore et al. (1975a) also administered $^{191}PtCl_4$ (25 µCi/animal) intravenously to 15 pregnant rats on day 18 of gestation to determine placental transfer after 24 h. High levels of ^{191}Pt radioactivity were found in the kidney (127 064 counts/g) and liver (43 375 counts/g), compared with 10 568 counts/g in the blood. Accumulation was also found in the placenta (27 750 counts/g). ^{191}Pt was detected in the 60 fetuses examined, but only at very low concentrations (an average of 432 counts/g). Thus, the placental barrier is crossed to a limited extent.

In contrast to the simple platinum salts, the diammine complexes such as cisplatin (see footnote in section 1.2) are excreted primarily in the urine. In mice, Hoeschele & Van Camp (1972) found about 90% of the intraperitoneally injected dose in the urine within five days. Little or no excretion occurred via the faeces. A high urinary recovery was also observed in rats and dogs (Hoeschele & Van Camp, 1972; Lange et al., 1972; Litterst et al., 1976a,b, 1979; Cvitkovic et al., 1977).

The excretion of both cis- and trans-diamminedichloroplatinum(II) follows a biphasic pattern with a fast initial α-phase and a second slow β-phase. The variation in the plasma half-lives is due to species differences and variations in dose, route of administration, time points analysed, and analytical method used (Litterst et al., 1979). The extremely rapid α-phase accounts for early, high levels of platinum in kidney, liver, skin, bone, ovary, and uterus. The prolonged β-phase results in detectable urine platinum concentrations 30 days after a single dose.

For both the simple platinum salts and cisplatin complexes, an initial rapid clearance is followed by a prolonged clearance phase during the remaining post-exposure period, and there is no evidence for markedly different retention profiles between these two groups of platinum compounds (Rosenberg, 1980).

All animal species studied show a similar organ distribution pattern for cisplatin. An initial distribution to nearly all tissues is followed by accumulation in the first hour mainly in kidney, liver, muscle, and skin. By the end of the first day, plasma levels decrease rapidly and there are elevated platinum levels in numerous other tissues (Litterst et al., 1979).

Cisplatin is extensively bound to plasma proteins; 90% of it may be bound 2 h after an intravenous injection. The bound portion is no longer cytotoxic (Safirstein et al., 1983; Sternson et al., 1984). In addition to its reactivity with plasma protein, renal excretion leads to a very low concentration of free cisplatin in the plasma and to a rapid accumulation in the kidney. Due to the presence of high chloride ion concentrations, cisplatin is relatively stable in extracellular fluids (see also section 7.6), which explains why it is excreted mainly in the unchanged form in human and rat urine (Safirstein et al., 1983).

7. EFFECTS ON LABORATORY MAMMALS AND *IN VITRO* TEST SYSTEMS

7.1 Single exposure

Acute toxicity data on platinum mainly relate to its coordination complexes, the chloroplatinates and ammines. Hofmeister (1882) was one of the first to test ammonium salts containing divalent and tetravalent platinum with various numbers of ammine ligands. He injected solutions of platinum complexes into the dorsal lymphatic sac of single frogs and subcutaneously into the dorsal skin of single rabbits. The symptoms observed included vomiting and diarrhoea with bloody stools and a "curare-like" action of the salts.

The acute toxicity of platinum depends considerably on the species of platinum involved (Table 14). Soluble platinum compounds are much more toxic. Hence, in the study of Holbrook (1976a) oral toxicity to rats decreased in the following order: $PtCl_4$ > $Pt(SO_4)_2 \cdot 4H_2O$ > $PtCl_2$ > PtO_2. For the two latter compounds no LD_{50} could be derived.

Signs of poisoning observed, for example, with $(NH_4)_2[PtCl_4]$, include hypokinesia, piloerection, diarrhoea, clonic convulsions, laboured respiration, and cyanosis (Degussa, 1989a).

Hexachloroplatinic acid is highly nephrotoxic in rats. After an intraperitoneal LD_{50} injection of 40-50 mg/kg, rats died of renal failure, hypocalcaemia, and hyperkalaemia. The necrotizing renal tubular lesions involved the entire renal cortex (Ward et al., 1976).

In its metallic state, platinum has an extremely low acute toxicity. Thus some alloys containing platinum are used in protheses. Fine dust particles of metallic platinum, 1-5 μm in diameter, orally administered to rats caused only slight necrotic changes in the gastrointestinal epithelium, granular dystrophy of hepatocytes, and swelling in the epithelium of the convoluted renal tubules (Roshchin et al., 1979, 1984). The highest dose given was not lethal. The dose was reported as "129 μA/kg" (25 167 μg per kg; personal communication from Prof. A.V. Roshchin to IPCS dated 3 April 1991).

Effects on Laboratory Mammals and In Vitro Test Systems

Due to the different absorption rates for platinum compounds, the route of administration also affects the toxicity, the intraperitoneal and intravenous routes leading to much higher toxicity than the oral route (Table 14).

Table 14. Acute toxicity of platinum and platinum compounds after oral (p.o.), intraperitoneal (i.p.), and intravenous (i.v.) administration to rats

Compound	Route	Sex[a]	LD_{50} (mg/kg)	Reference
PtO_2	p.o.	m	> 8000	Holbrook et al. (1976a,b)
$PtCl_2$	p.o.	m	> 2000	Holbrook et al. (1976a,b)
$PtCl_2$	p.o.	m	3423[b]	Roshchin et al. (1984)
$PtCl_2$	i.p.	m	670	Holbrook et al. (1976a,b)
$PtCl_4$	p.o.	m	240	Holbrook et al. (1976a,b)
$PtCl_4$	p.o.	m/f	276[b]	Roshchin et al. (1984)
$PtCl_4$	i.p.	m	38	Holbrook et al. (1976a,b)
$PtCl_4$	i.v.	m	26.2	Moore et al. (1975b)
$PtCl_4$	i.v.	m	41.4	Moore et al. (1975b)
$Pt(SO_4)_2 \cdot 4 H_2O$	p.o.	m	1010	Holbrook et al. (1976a,b)
$Pt(SO_4)_2 \cdot 4 H_2O$	i.p.	m	310[c]	Holbrook et al. (1976a,b)
$Pt(SO_4)_2 \cdot 4 H_2O$	i.p.	m	138-184[c]	Holbrook et al. (1976a,b)
$(NH_4)_2[PtCl_6]$	p.o.	m/f	195[b]	Roshchin et al. (1984)
$(NH_4)_2[PtCl_6]$	p.o.	m/f	≈ 200	Johnson Matthey (1978a)
$(NH_4)_2[PtCl_4]$	p.o.	m	212	Degussa (1989a)
$(NH_4)_2[PtCl_4]$	p.o.	f	125	Degussa (1989a)
$H_2[PtCl_6]$	i.p.	m	40-50	Ward et al. (1976)
$Na_2[PtCl_6]$	p.o.	m/f	25-50	Johnson Matthey (1978b)
$Na_2[Pt(OH)_6]$	p.o.	m/f	500-2000	Johnson Matthey (1978c)
$K_2[PtCl_4]$	p.o.	m/f	50-200	Johnson Matthey (1981a,b)
$K_2[Pt(CN)_4]$	p.o.	m/f	> 2000	Johnson Matthey (1977a)
$[Pt(NH_3)_4]Cl_2$	p.o.	m/f	> 15 000	Johnson Matthey (1977b)
$[Pt(NO_2)_2(NH_3)_2]$	p.o.	m	≈ 5000	Degussa (1989b)
$[Pt(NO_2)_2(NH_3)_2]$	p.o.	f	> 5110	Degussa (1989b)
$[Pt(C_5H_7O_2)_2]$	p.o.	m/f	> 500	Johnson Matthey (1976a)
cis-$[PtCl_2(NH_3)_2]$[d]	p.o.	m/f	≈ 20	Johnson Matthey (1977c)
cis-$[PtCl_2(NH_3)_2]$[d]	i.p.	m	12	Kociba & Sleight (1971)
cis-$[PtCl_2(NH_3)_2]$[d]	i.p.	m	7.7	Ward & Fauvie (1976)
cis-$[PtCl_2(NH_3)_2]$[d]	i.v.	m	7.4	Ward et al. (1976)
trans-$[PtCl_2(NH_3)_2]$	p.o.	m/f	> 5110	Degussa (1989c)

[a] m = male; f = female
[b] Calculated from the original values given as mg A/kg (= milligramme atom/kg)
[c] Results from two different laboratories
[d] See footnote in section 1.2

7.2 Short-term exposure

Holbrook et al. (1975) conducted repeated-dose oral toxicity studies on male Sprague-Dawley rats. The soluble salts $PtCl_4$ and $Pt(SO_4)_2 \cdot 4H_2O$ were added to the drinking-water, which was consumed *ad libitum*. Within the observation period of 4 weeks, a concentration of 0.54 mmol/litre (182 mg $PtCl_4$/litre or 248 mg $Pt(SO_4)_2 \cdot 4H_2O$ per litre) did not affect the normal weight gain. A 3-fold increase in the platinum concentration to 1.63 mmol/litre reduced the weight gain by about 20% during the first week only; this paralleled a 20% decrease in feed and fluid consumption. The dietary administration of $PtCl_4$ at concentrations of 0.5 mmol/litre for approximately 30 days or 1.6 mmol/litre for 8 days (169 and 539 mg/litre, respectively) did not affect the weights of any of the five organs investigated, i.e. liver, kidney, spleen, heart, and testes. Similarly, the administration of 1.6 mmol per litre of $Pt(SO_4)_2 \cdot 4H_2O$ (734 mg/litre) for 8-9 days did not significantly affect organ weights. Total platinum intake for each of these three experimental conditions was approximately 50 mg per rat. When 1.6 mmol $PtCl_4$/litre (539 mg/litre) was given for about 30 days (total intake of about 250 mg Pt per rat), the kidney weight increased by about 6-10%. No effects on the level of microsomal protein or the activities of aniline hydroxylase and aminopyrine demethylase in liver microsomes were found (Holbrook, 1976b).

7.3 Skin and eye irritation; skin and respiratory sensitization

7.3.1 Skin irritation

The dermal irritancy of several platinum compounds was tested on albino rabbits using comparable procedures and evaluation criteria. Platinum test materials were spread on abraded and intact skin sites, located dorsolaterally on the animals' trunks. The skin reactions were evaluated after 24, 48, and 72 h, and are summarized in Table 15.

7.3.2 Eye irritation

Summarized data on eye irritation are presented in Table 15. All tested platinum salts were either corrosive or irritating to varying degrees.

Table 15. Skin and eye irritation by platinum compounds[a]

Compound	Primary irritation score	Skin irritation test[b] Classification	Reference	Eye irritation test[c] Classification	Reference
PtO_2	0	non-irritant	Campbell et al. (1975)		
$PtCl_2$	0.4[d]	non-irritant	Campbell et al. (1975)		
$PtCl_4$	2.2[e]	irritant	Campbell et al. (1975)		
$(NH_4)_2[PtCl_6]$	1.3	mild irritant	Johnson Matthey (1978d)		
$(NH_4)_2[PtCl_4]$	2.7	slight irritant	Degussa (1988a)	corrosive	Degussa (1988b)
$Na_2[PtCl_6]$	0.5	mild irritant	Johnson Matthey (1978e)	irritant	Johnson Matthey (1978f)
$Na_2[Pt(OH)_6]$	5.4	severe irritant	Johnson Matthey (1978g)		
$K_2[PtCl_4]$	0[f]	non-irritant	Johnson Matthey (1981c)	irritant	Johnson Matthey (1981d)
$K_2[Pt(CN)_4]$	0.3	mild irritant	Johnson Matthey (1977d)	irritant	Johnson Matthey (1978h)
$[Pt(NH_3)_4]Cl_2$	2.8	moderate irritant	Johnson Matthey (1977e)	strongly irritant	Johnson Matthey (1977f)
$[Pt(NO_2)_2(NH_3)_2]$	0	non-irritant	Degussa (1989d)	severely irritant	Degussa (1989e)
$[Pt(C_5H_7O_2)_2]$	0	non-irritant	Johnson Matthey (1976b)	mildly irritant	Johnson Matthey (1976c)
cis-$[PtCl_2(NH_3)_2]$	0.13	mild irritant	Johnson Matthey (1977g)	severely irritant (toxic)	Johnson Matthey (1977h)
$trans$-$[PtCl_2(NH_3)_2]$	0	non-irritant	Degussa (1988c)	corrosive	Degussa (1988d)

[a] Adapted from Bradford (1988)
[b] The skin tests (patch tests on albino rabbits) were carried out according to the US Federal Register 1973 Skin Test (24 h-contact) (Johnson Matthey) or according to OECD Test Guideline No. 404 (4-h contact) (Degussa). The method used by Campbell et al. (1975) is comparable to these tests.
[c] The eye irritation tests on albino rabbits were carried out according to the US Federal Register 1973 Eye Test (Johnson Matthey) or according to OECD Guideline No. 405 (Degussa).
[d] Average score from 0.2 (intact skin) and 0.6 (abraded skin); a score of 0-0.9 was considered as "non-irritant" by the authors.
[e] Average score from 1.8 (intact skin) and 2.6 (abraded skin); a score of 2 was considered as "irritant" by the authors.
[f] Using OECD Test Guideline No. 404 (4-h contact).

7.3.3 Skin sensitization

In a study by Kolpakova & Kolpakov (1983), platinum hydrochlorides administered intravenously to rabbits in repeated doses induced sensitization confirmed by the basophil degranulation test, neutrophil damage index, leucocyte agglomeration, neutrophil alteration, and the drop skin and skin fenestra tests. These data are unusual and have not been confirmed in other studies.

Taubler (1977) injected rabbits, guinea-pigs, and mice subcutaneously and intravenously with $PtSO_4$ (0.05-0.3 mg/litre with and without NH_4Cl) three times a week for 4 weeks. No induction of an allergic state was found, as measured by skin tests (guinea-pigs and rabbits), passive transfer, and footpad tests (mice). Administration of platinum-egg-albumin complex also failed to sensitize the experimental animals.

In a study by Murdoch & Pepys (1985), rats were immunized with ovalbumin-platinum. Sera of the animals which were positive in the passive cutaneous anaphylaxis (PCA) test and a radioallergosorbent test (RAST) were pooled and used for PCA tests with other platinum salts having differing ligands. A significant cross-reactivity between ammonium tetrachloroplatinate(II), ammonium hexachloroplatinate(IV), and the conjugated tetrachloroplatinate was observed. There was very limited or no cross-reactivity with the compounds cesium trichloronitroplatinate(II), cis-diamminedichloroplatinum(II), potassium tetracyanoplatinate(II), and tetraammineplatinum(II) chloride.

7.3.4 Skin and respiratory sensitization

Biagini et al. (1983) exposed two groups of Cynomolgus monkeys *(Macaca fasicularis)* to disodium hexachloroplatinate, $Na_2[PtCl_6]$, by nose-only inhalation of 200 and 2000 $\mu g/m^3$, 4h/day, biweekly for 12 weeks. Another group was exposed percutaneously to the salt (20 mg/ml) applied biweekly to an open patch area in the intrascapular region. Two weeks after termination of exposure, bronchoprovocation challenges with $Na_2[PtCl_6]$ and pulmonary function tests were performed. Percutaneous application did not affect post-challenge pulmonary function. The 200 $\mu g/m^3$ group showed significantly greater pulmonary deficits as compared to control animals.

Average pulmonary flow resistance (R_L) was significantly increased, while forced expiratory volume in 0.5 seconds, corrected for vital capacity ($FEV_{0.5}/FVC$), was decreased. No dermal hypersensitivity was observed. The question of whether the observed pulmonary hyper-reactivity is due to a superpharmacological, irritant, local immune, or combination mechanism is unresolved. The absence of hyper-reactivity in the 2000-$\mu g/m^3$ group suggests a possible pulmonary tolerance mechanism, tachyphylaxis, or delay in the development of symptoms at higher sensitization concentrations.

7.3.5 Respiratory sensitization

In a 12-week inhalation experiment with Cynomolgus monkeys exposed to either ammonium hexachloroplatinate (200 $\mu g/m^3$) or ozone (2000 $\mu g/m^3$; 1 ppm) alone or as a combination of both, Biagini et al. (1986) found significant allergic platinum dermal hypersensitivity, based on concentrations necessary to give a positive test, and pulmonary hyper-reactivity only with concomitant exposure to ozone. Inflammation, epithelial damage, cell recruitment, and modifications of cellular tight junctions caused by ozone may increase the penetration of platinum into the pulmonary epithelium and subepithelial tissue. This could lead to increased protein binding sites or absorption of the platinum salts and finally to the development of pulmonary hyper-reactivity and allergic sensitization (Biagini et al., 1986).

7.3.6 Sensitization by other routes

Murdoch & Pepys (1984) investigated the immunological responses to complex platinum salts in the female hooded Lister rat, a strain that produces high and consistent levels of circulating IgE when immunized with low doses of antigen together with *Bordetella pertussis* adjuvant, and that reacts with enhanced synthesis of IgE upon secondary boosting. Sensitization with the free salt of ammonium tetrachloroplatinate, $(NH_4)_2[PtCl_4]$, was attempted via the intraperitoneal, intramuscular, intradermal, subcutaneous, intratracheal, and footpad routes over a wide range of doses (1 to 1000 μg). Both *B. pertussis* and/or aluminium hydroxide gel were added as adjuvants. As shown by direct skin testing using the PCA or RAST methods, no sensitization was achieved. However, sensitization was obtained by intraperitoneal injection of the platinum salt conjugated to ovalbumin (OVA). Antibodies were produced to Pt-OVA and to OVA alone. Specific sensitization was demonstrated both by PCA challenge with Pt-BSA (no positive PCA reactions were seen with BSA alone) and by positive RAST, demonstrated by RAST inhibition techniques with a Pt-BSA conjugate.

7.4 Reproductive toxicity, embryotoxicity, and teratogenicity

Only limited experimental data concerning the effects of platinum on reproduction, embryotoxicity, and teratogenicity are available. D'Agostino et al. (1984) studied the embryotoxic effects of platinum compounds in Swiss ICR mice. Single doses of either $Pt(SO_4)_2.4H_2O$ or $Na_2[PtCl_6].6H_2O$ were administered intragastrically or subcutaneously, respectively, on the 7th and 12th day of gestation. The pups were cross-fostered to treated or untreated dams at birth and were culled to three animals of each sex per litter. In the $Pt(SO_4)_2$ study, the LD_1 dose of 200 mg Pt/kg caused a reduced offspring weight from day 8 to day 45 postpartum. The major effect of disodium hexachloroplatinate (20 mg Pt/kg) was a reduced activity level exhibited by the offspring of dams exposed on the 12th day of gestation. The general activity was quantified on an activity field consisting of concentric circles. The number of lines crossed during 1 min comprised the activity score. On days 60-65 postpartum, open-field behaviour (ambulation and rearing), rotarod performance, and passive avoidance learning were investigated in the adult offspring. No effects were found after administration on the 7th day, but administration on day 12 of gestation had significant behavioural effects.

Solid platinum, wire or foil, is considered to be biologically inert and adverse effects on implantation are probable due to the physical presence of a foreign object in the uterus (Barlow & Sullivan, 1982).

Kraft et al. (1978) reported normal fertility in male rabbits with open tube gold/platinum devices inserted into the vas deferens. There was an initial decrease in sperm count and motility, but these parameters returned to normal after three weeks. At 117-426 days after insertion, 7 out of 9 animals were fertile in numerous matings.

Effects on human sperm motility were investigated by Kesseru & Leon (1974). Fresh sperm were incubated for up to 5 h in the presence of strips of platinum or other metals. Motility after 2 and 5 h was 60 and 30%, respectively, compared to 10 and 0% for copper, 40 and 7% for silver, and 90 and 65% for gold.

Platinum wire inserted into the uterus of rats was reported to reduce the implantation of fertilized ova. An 83% reduction in the number of implantation sites in the affected uterine horn, compared to the unoperated horn, was found in rats unilaterally implanted on day 3 (Chang et al., 1970). Chang & Tatum (1975) found no effect on embryonic or fetal survival if platinum wire was inserted after implantation on day 6. Tobert & Davies (1977) showed a 37% reduction in the number of implanting ova in the uteri of rabbits containing platinum foil.

7.5 Mutagenicity and related end-points

The genotoxic effects of platinum compounds have been investigated in bacterial systems, mammalian cell cultures and *in vitro* studies.

In bacteria many of the tested platinum compounds were moderately mutagenic. Cisplatin and some of its analogues showed the greatest mutagenic potential; other platinum compounds were less mutagenic.

In Ames tests, nearly all using the test strains *Salmonella typhimurium* TA98 and TA100, positive results were reported (Lecointe et al., 1977; Andersen, 1979; Suraikina et al., 1979; Life Science Research, 1980a; Kanematsu et al., 1980). With $[Pt(NH_3)_4]Cl_2$, mutagenic potential was observed in strain TA1537 with and without S-9 metabolic activation (Life Science Research, 1980a).

The induction of reverse mutations in the plasmid-carrying strains TA98 and TA100 indicated base-pair substitution and frame-shift mutations (Lecointe et al., 1977; Suraikina et al., 1979; Kanematsu et al., 1980).

$(NH_4)_2[PtCl_6]$ but not $PtCl_4$ caused base-change mutation in *Escherichia coli* B/r WP2 (Kanematsu et al., 1980).

The growth of a Rec strain of *Bacillus subtilis* was significantly inhibited by $(NH_4)_2[PtCl_6]$ (0.1 mol/litre), $H_2[PtCl_6]$ (0.01 mol/litre), and $PtCl_4$ (0.01 mol/litre) (Kanematsu et al., 1980).

In a mutagenic test with the mouse lymphoma cell line L 5178Y, cisplatin, transplatin, and $PtCl_4$ produced significantly higher mutation frequencies than occurred in the controls, but $[Pt(NO_2)_2](NH_3)_2$ and $PtCl_2$ did not (Sandhu, 1979).

Cellular resistance to the toxic effects of two platinum complexes was introduced into Chinese hamster ovary (CHO) cells by continuous exposure to $K_2[PtCl_6]$ and $Pt(SO_4)_2$ for 5 and 4 months, respectively. These cell lines had resistant phenotypes stable for at least 55 population doublings in the absence of a platinum compound. The induced resistances were interpreted by the authors to be a result of mutation and selection (Smith et al., 1984).

In a micronucleus test in mice involving oral administration of $[Pt(NH_3)_4]Cl_2$, no significant increase in the incidence of micronucleated polychromatic erythrocytes was found. Additionally, $[Pt(NH_3)_4]Cl_2$ did not markedly inhibit bone marrow cell division at any level (Life Science Research, 1980b). Also, no evidence of induced chromosomal damage leading to micronucleus formation in polychromatic erythrocytes was observed after oral administration of $K_2[PtCl_4]$ in mice (Life Science Research, 1981a).

No significant increase in the incidence of aberrant metaphases was found in bone marrow cells after subacute oral administration of $[Pt(NH_3)_4]Cl_2$ or $K_2[PtCl_4]$ to Chinese hamsters (Life Science Research, 1981b, 1982).

$K_2[PtCl_4]$ and $[Pt(NH_3)_4]Cl_2$ induced no increase in the frequency of sex-linked recessive lethal mutations in *Drosophila melanogaster* (Life Science Research, 1980c, 1981c).

In a structure-mutagenicity study with the CHO:HGPRT-system, cis-[Pt(NH$_3$)$_2$Cl$_2$] was the most potent of six platinum compounds tested. Based on the slope of the mutation induction curve, the approximate relative mutagenic activity of cis-[Pt(NH$_3$)$_2$Cl$_2$], K[Pt(NH$_3$)Cl$_3$], and [Pt(NH$_3$)$_3$Cl]Cl was 100:9:0.3. The mutation frequency for K$_2$[PtCl$_4$] and $trans$-[Pt(NH$_3$)$_2$Cl$_2$] was related to the concentration used, but was not much greater than the maximum spontaneous mutation frequency. No mutagenic activity was observed for [Pt(NH$_3$)$_4$]Cl$_2$. The relative cytotoxicity of the tested compounds was similar. Additionally, it was found that cis- and $trans$-[Pt(NH$_3$)$_2$Cl$_2$] bind to DNA after entering the cell, but the relative mutagenicities are not a consequence of different initial levels of DNA binding (Johnson et al., 1980).

Dose-dependent forward mutations were induced by PtCl$_4$ to 8-azaguanine resistance (8-AG^R/HGPRT locus) in Chinese hamster ovary (CHO-S) cells. In addition there was an increased dose-related frequency of CHO-AUXB1 reversion (Taylor et al., 1979)

Cisplatin, which is not reviewed in detail in this document (see footnote in section 1.2), induces structural chromosomal aberrations and sister chromatid exchanges in cells of rodents treated in $vivo$, chromosomal aberrations, micronuclei, and sister chromatid exchanges in both human and rodent cells in $vitro$, and mutation and DNA damage in rodent cells in $vitro$. Cisplatin is also mutagenic in $Drosophila$, fungal, and bacterial test systems (IARC, 1987a).

7.6 Carcinogenicity and anticarcinogenicity

No experimental data are available on the carcinogenicity of platinum and platinum compounds except for cisplatin (see footnote in section 1.2). IARC (1987b) considered sufficient the evidence for the carcinogenicity of cisplatin for animals (see chapter 13).

Cisplatin and its analogues, however, are exceptional compared to the other platinum compounds. This is reflected by the unique mechanism for their anti-neoplastic activity demonstrated in in $vitro$ studies (Rosenberg, 1980, 1985). At low doses cisplatin produces specific inhibition of DNA synthesis (but not of RNA and protein synthesis) by causing DNA lesions such as monofunctional adducts, bifunctional binding to a single base moiety, and DNA cross-links of inter- and intrastrand types (Harder & Rosenberg, 1970; Howle & Gale, 1970). There is sufficient evidence that the DNA cross-links are responsible for cellular toxicity, but not for anti-tumour activity. For the latter, another observation probably plays the decisive role; only the cis isomer forms a closed ring chelate of the aquated cisplatin with guanine at a certain position of guanine. Thus, intrastrand DNA cross-linking is considered to be the most important reason for anti-tumour activity. It appears that, due to the cisplatin-induced DNA cross-links, the replication of DNA is impaired in cancer cells, while in normal cells the cisplatin lesions on guanine are repaired before replication (Rosenberg, 1985; Pinto & Lippard, 1985).

The high chloride concentration of the extracellular fluid (112 mmol/litre) is sufficient to limit the substitution of water ligands for chloride. However, within the cell the platinum complex is exposed to low chloride concentrations (4.4 mmol/litre) and hydrolysis of the chloride leaving groups can occur (Rosenberg, 1975), a process that has been shown to accelerate the rate of reaction of platinum with DNA (Johnson et al., 1980) and to increase its toxicity (Litterst, 1981). This hydration provides the only known activation process required for cisplatin to react with molecules in the cell, and metabolic activation is not required (Rosenberg, 1980). The binding of cisplatin with plasma proteins, on the other hand, is not inhibited by chloride and presumably involves a different mechanism (De Conti et al., 1973), such as the strong electrophiles on proteins (Cleare, 1977a).

7.7 Other special studies

7.7.1 Effects on alveolar macrophages

Rabbit alveolar macrophages exposed to the water-soluble platinum(IV) chloride at a concentration of 0.4 mmol/litre (78 mg Pt/litre) for a period of 20 h exhibited a 50% reduction in macrophage viability. A reduction in phagocytic activity and a decrease in total cellular adenosine triphosphate to 50% of the value in control macrophages was observed at 0.21 and 0.25 mmol/litre (41 and 48 mg Pt/litre). Platinum(IV) oxide (PtO_2) did not dissolve in the culture medium and, hence, was ineffective at concentrations as high as 500 mg/litre (Waters et al., 1975).

7.7.2 Non-allergic mediator release

Investigations in guinea-pigs, rats, and dogs showed an increase in bronchomotility and histamine release after intravenous treatment with disodium chloroplatinate, $Na_2[PtCl_6]$ (Saindelle & Ruff, 1969; Parrot et al., 1969).

Saindelle & Ruff (1969) noticed dyspnoea one minute after an intravenous injection of disodium chloroplatinate (20 mg/kg) into guinea-pigs. Within 5 min an intense attack of asthma occurred resulting in death. Histamine release occurred following the injection and the blood histamine level was greatly increased. The injection of a smaller dose (1-2 mg/kg) resulted in bronchospasm comparable to that caused by 3 µg/kg of histamine. Repeated injections of histamine caused reproducible changes in bronchial motility, whereas the platinum compound caused tachyphylaxis.

The intense breathing difficulties observed in these studies were presumably due to non-allergenic histamine release. This nonspecific histamine release has complicated the interpretation of both animal and human studies with respect to the conclusion of allergic sensitization.

7.7.3 Effects on mitochondrial function

No pronounced effects of platinum on the mitochondrial function of liver, heart, lung or kidney cells were observed in an *in vitro* test on succinate-stimulated O_2 uptake 24 and 48 h after intragastric administration of 40 and 80 µmol $K_2[PtCl_6]$/litre (19 and 38 mg/litre) to Sprague-Dawley rats (Michael et al., 1976).

7.7.4 Effects on the nervous system

The open field behaviour of adult Swiss mice has been found to be influenced by platinum salts administered intragastrically in the form of a single dose or of repeated doses. A single dose of $Pt(SO_4)_2$ at the LD_{25} level (213 mg Pt/kg body weight) depressed ambulation significantly and rearing marginally. For ambulation, this pattern persisted from 4 h to 7 days after administration, although the effect was most obvious at 4 h. Repeated doses of the same salt at the LD_1 level (up to 10 doses of 109 mg Pt/kg every 72 h) caused a marginal depression of activity and exploratory behaviour (Lown et al., 1980). Also, a single dose of $Na_2[PtCl_6]$ depressed ambulation significantly (Massaro et al., 1981).

During the course of a reproduction study, behavioural effects were observed in the offspring of mice treated with sodium hexachloroplatinate (see section 7.4; D'Agostino et al., 1984).

7.7.5 Side effects of cisplatin and its analogues[a]

As discussed in section 8.3, the therapeutic use of cisplatin in humans can be accompanied by several toxic side-effects. In animal studies, only some of which are presented in this monograph, similar effects were observed.

Ward et al. (1976) investigated the nephrotoxicity of cisplatin and its analogues in male F-344 (Fischer CDF) rats. An intravenous LD_{50} dose of cisplatin (7.4 mg Pt/kg body weight) caused an increase in the blood urea nitrogen and creatinine levels reaching a peak on days 4 to 5. Diarrhoea developed by day 3. Necrotizing enteritis of the small intestine, caecum, and colon, cellular depletion of bone marrow and thymus, and acute degenerative and necrotizing renal tubular lesions also occurred (Ward et al., 1976).

Oxoplatinum (cis-dichlorodiammine-trans-dihydroxyplatinum(IV)) also caused marked nephrotoxicity after intravenous administration (20 mg/kg as a single dose) to rats. However, another cisplatin analogue, CBDCA (cis-diammine-1,1-cyclobutane dicarboxylate platinum(II)), did not result in significant changes in renal function parameters (Laznickova et al., 1989).

---

[a] See footnote in section 1.2.

Cisplatin has been found to cause bone marrow suppression. The surviving fraction of haemopoietic bone-marrow system cells in mice decreased from 1 to 0.03 after treatment with an LD_{50} dose of cisplatin (Lelieveld et al., 1984). A 36-53% decrease in lymphocyte and granulocyte counts was observed in mouse bone marrow after intraperitoneal treatment with 5 mg cisplatin/kg (Bodenner et al., 1986).

Cisplatin administered intraperitoneally (6 mg/kg) has been shown to affect gastric emptying in rats. There was a large increase in the weight of the stomach due to retained food (Whitehouse & Garrett, 1984).

In dogs, cisplatin given intravenously as a dose of 2 mg/kg resulted in a complete interruption of interdigestive myoelectric activity of the gastric antrum, duodenum, and jejunum (Chey et al., 1988).

Ototoxicity has been demonstrated in guinea-pigs. A cisplatin dose of 1.5 mg/kg administered intraperitoneally once a day caused hearing loss beginning at about the ninth day of administration (Hoeve et al., 1987).

7.8 Factors modifying toxicity

Physiological levels of selenium administered simultaneously with food to mice markedly depressed the acute toxicity of some platinum salts by forming inert complexes of high relative molecular mass in the presence of protein (Imura, 1986).

8. EFFECTS ON HUMANS

8.1 General population exposure

8.1.1 Acute toxicity - poisoning

Except for one case of poisoning in 1896 (Hardman & Wright, 1896), no acute poisoning cases have been reported.

8.1.2 Effects of exposure to platinum emitted from automobile catalysts

An immunological study conducted by Cleare (1977b) addressed the question of whether the emitted platinum is allergenic. He investigated the response of individuals, who were highly sensitive to platinum salt (skin test positive at low platinum salt concentrations), to extracts of particulate exhaust samples. The total platinum content at the highest concentration was more than 5 μg/ml, which would normally be sufficient to elicit a response. Five extracts tested on three people, using the skin prick test, did not elicit a positive response.

8.2 Occupational exposure

The occupational hazards of platinum are principally confined to some halogenated complex platinum salts (Rosner & Merget, 1990).

8.2.1 Case reports and cross-sectional studies

A report of health problems arising from occupational exposure to platinum was produced by Karasek & Karasek (1911). They studied workers in photographic studios in Chicago handling photographic paper treated with complex platinum salts. The symptoms observed in eight workers were pronounced irritation of the nose and throat causing violent sneezing and coughing, together with difficulties in breathing.

Hunter et al. (1945) conducted environmental and clinical studies on workers in four British platinum refineries. Out of 91 workers exposed to complex salts of platinum, 52 showed symptoms starting with repeated sneezing and rhinorrhoea, followed by tightness of the chest, shortness of breath, cyanosis, and wheezing. Scaly erythematous dermatitis of hands and forearms, sometimes also affecting the face and neck, and urticaria were observed in 13 workers. The respiratory tract symptoms persisted during working hours and for about one hour after leaving the factory. The latency period from the first contact with platinum to the occurrence of the first symptoms varied from a few months to six years. Once skin and respiratory tract sensitization was established, symptoms tended to become worse as long as the workers were exposed to platinum salts.

In the USA, Roberts (1951) studied 21 employees of a platinum refinery for five years. All workers showed some form of platinum-related disease for which Roberts (1951) introduced the term "platinosis". According to his classification of this occupational disease, 40% of the employees did not have typical symptoms but exhibited the same inflammatory changes in the conjunctivae and the mucous membranes of the upper respiratory tract as were seen in the 60% of the workers with definite symptoms.

These observations have been confirmed by other investigators. The term "platinosis" is no longer used, as it implies a chronic fibrosing lung disease. This had been assumed by Roberts (1951) but has not been reported elsewhere. The terms "platinum salt allergy" (Schultze-Werninghaus et al., 1978), "platinum salt sensitivity" (Linnett, 1987), "allergy to platinum compounds containing reactive halogen ligands" (Hughes, 1980) and "platinum salt hypersensitivity" (PSH) have been used, with the latter being preferred.

The symptoms typical of platinum salt sensitization (Roshchin et al., 1979; Health and Safety Executive, 1983; Brooks et al., 1990) include watering of the eyes, sneezing, tightness of the chest, wheezing, breathlessness, coughing, eczematous and urticarial skin lesions, signs of mucous membrane inflammation.

In earlier studies, the prevalence of allergic symptoms due to platinum salt exposure was as high as 80% (Table 16) although this was not consistently confirmed by skin testing. Estimated workplace exposure concentrations ranged from 0.9 to 1700 µg Pt/m^3 (Hunter et al., 1945). However, due to analytical deficiencies, these data do not allow the quantification of these exposure situations. It can be assumed that occupational exposure is much lower today due to improved engineering control and occupational exposure limits. Airborne dust analyses in a platinum refinery revealed levels between 0.08 and 0.1 µg/m^3 in the separation department; in other areas the measurements were all below 0.05 µg/m^3 (Bolm-Audorff et al., 1988) or below 0.08 µg/m^3 (Merget et al., 1988). Work-related symptoms were reported in 8 to 23% of workers exposed to these concentrations (Table 16). The risk of developing platinum salt hypersensitivity seems to be correlated with the intensity of exposure. In the surveys of Bolm-Audorff et al. (1988) and Merget et al. (1988), the highest rates of prevalence occurred in the groups exposed to the highest concentration.

Table 16. Prevalence of symptoms and positive skin tests in refinery workers exposed to platinum salts

Total workers[a]	Workers with symptoms[b]	Prevalence of symptoms (%)[c]	Reference
91 (16)	52 (4)	57 (25)	Hunter et al. (1945)
20 (19)	12 (8)	60 (42)	Roberts (1951)
15 (nd)	12 (nd)	80 (nd)	Massmann & Opitz (1954)
51 (nd)	35 (nd)	69 (nd)	Hebert (1966)
107 (107)	31[d] 47[e] (15)	29[d] 44[e] (14)	Biagini et al. (1985); Brooks et al. (1990)
65 (64)	15 (12)	23 (19)	Bolm-Audorff et al. (1988)
24 (20)	2 (4)	8 (20)	Merget et al. (1988)

[a] Values in parentheses are numbers of skin-tested workers
[b] Values in parentheses are numbers of workers with a positive skin test
[c] Values in parentheses give the prevalence of positive skin tests as a percentage of skin-tested workers
[d] Workers with upper respiratory tract symptoms
[e] Workers with lower respiratory tract symptoms
nd = not determined

Effects on Humans

8.2.2 Allergenicity of platinum and platinum compounds

Metallic platinum seems to be non-allergenic. With the exception of a single reported case of alleged contact dermatitis from a "platinum" ring (Sheard, 1955), no allergic reactions have been reported.

Halogenated platinum salts are among the most potent sensitizers. The compounds mainly responsible for platinum sensitization are hexachloroplatinic acid, $H_2[PtCl_6]$, and the chlorinated salts such as ammonium hexachloroplatinate, $(NH_4)_2[PtCl_6]$, potassium tetrachloroplatinate, $K_2[PtCl_4]$, potassium hexachloroplatinate, $K_2[PtCl_6]$, and sodium tetrachloroplatinate, $Na_2[PtCl_4]$. Cleare et al. (1976) investigated the allergenic potency of platinum complexes by means of skin prick tests on platinum refinery workers who were known to be sensitive to hexachloroplatinate. Their results suggest that platinum allergy is confined to a small group of charged compounds that contain reactive ligand systems, the most effective of which are chloride ligands. The allergic response generally increases with increasing number of chlorine atoms, as demonstrated by the following sequence of potency:

$(NH_4)_2[PtCl_6] \approx (NH_4)_2[PtCl_4] > Cs_2[Pt(NO_2)Cl_3] > Cs_2[Pt(NO_2)_2Cl_2] > Cs_2[Pt(NO_2)_3Cl]$

Ionic platinum compounds containing bromide or iodide are also allergenic, but are less effective.

Non-halogenated complexes such as $[Pt\{(NH_2)_2CS\}_4]Cl_2$, $K_2[Pt(NO_2)_4]$, and $[Pt(NH_3)_4]Cl_2$, and neutral complexes such as cisplatin, cis-$[PtCl_2(NH_3)_2]$, are not allergenic, probably because they do not react with proteins to form a complete antigen. Anaphylactic shock reactions observed after the intravenous administration of relatively high doses of cisplatin (Khan et al., 1975; Von Hoff et al., 1979) were probably caused by contamination with the potent hexa- or tetrachloroplatinate (Pepys, 1983).

8.2.3 Clinical manifestations

The latency period from the first exposure to platinum salts to the occurrence of the first symptoms usually varies between three months and three years (Parrot et al., 1969; Schultze-Werninghaus et al., 1978; Ruff et al., 1979; Biagini et al., 1985), but is sometimes only a few weeks (Roberts, 1951; Hughes, 1980; Merget et al., 1988).

The dermatitis observed in the past (Roberts, 1951) is believed to have been mainly of a primary irritant nature following exposure to strong acids and alkalis. True contact dermatitis (i.e. allergic) is extremely rare. However, contact urticaria is seen in sensitized people following splashes with platinum salts and in some instances this is the first indication of sensitization (Hughes, 1980).

The symptoms usually worsen with increasing duration of exposure but generally disappear when the subject is removed from exposure. The latter was shown by the follow-up study carried out on platinum refinery workers who had to cease work with platinum salts because of sensitization (Newman-Taylor, 1981). This study found no evidence of long-term effects when workers giving a positive skin prick test and showing symptoms of platinum sensitization were removed immediately from contact with platinum salts. However, Schultze-Werninghaus et al. (1989) reported that after long duration exposure following sensitization individuals may never become completely free of symptoms. Similarly, Biagini et al. (1985) demonstrated the existence of positive platinum skin tests at very low concentrations in workers who had been free of occupational platinum exposure for periods of up to four years.

8.2.4 Immunological mechanism and diagnosis

The clinical manifestations of soluble platinum salt allergy reflect a true allergic response based on the following clinical criteria (Hughes, 1980; Biagini et al., 1985, 1986; Merget et al., 1988; Schultze-Werninghaus et al., 1989):

- the appearance of sensitivity is preceded by a symptomless exposure;
- not all exposed individuals become sensitized;
- the affected individuals become increasingly sensitive to platinum and react even at very low levels of exposure;

- negative prick test results are obtained in atopic and non-atopic controls.

The mechanism of platinum salt allergy appears to be a Type I (IgE mediated) response. The possibility of the formation of IgE antibodies to platinum chloride complexes in sensitized individuals has been assumed on the grounds of allergy and serological tests. It is believed that platinum salts of low relative molecular mass act as haptens combining with serum proteins to form the complete antigen. However, the actual immunological mechanism has not yet been defined (Zachgo et al., 1985).

It has been demonstrated that platinum(II) reacts with the sulfur atoms in the six methionine groups in human serum albumin (HSA) and that methionine 123 is the primary binding site (Grootveld, 1985).

Skin prick tests with freshly prepared solutions of soluble platinum complexes appear to provide reproducible, reliable, reasonably sensitive, and highly specific biological monitors of allergenicity (Cleare et al., 1976; Dally et al., 1980). The compounds used for routine screening of exposed workers are $(NH_4)_2[PtCl_6]$, $Na_2[PtCl_6]$, and $Na_2[PtCl_4]$. After sensitization due to previous exposure, prick testing with concentrations of the platinum compound in the range of 10^{-3} to 10^{-9} g/ml will produce immediate weal and flare reactions (Pepys et al., 1972; Pickering, 1972; Hughes, 1980; Gallagher et al., 1982; Biagini et al., 1985; Boggs, 1985; Jacobs, 1987; Linnett, 1987; Murdoch & Pepys, 1987; Schultze-Werninghaus et al., 1989). At these concentrations, nonspecific skin reactions were not found in atopic or non-atopic controls (Pepys et al., 1972; Murdoch & Pepys, 1987; Merget et al., 1988).

Passive transfer of immediate reactivity to intracutaneous tests in humans was demonstrated in the Prausnitz-Küstner test by Freedman & Krupey (1968). Schultze-Werninghaus et al. (1978) observed positive reactions in passive cutaneous anaphylaxis (PCA) in monkeys with serum from a platinum refinery worker. Similar tests were performed by Pepys et al. (1979) and Biagini et al., (1985). The results, however, were inconsistent, because positive as well as negative Prausnitz-Küstner prick test or PCA reactions were elicited in human recipients or in monkeys, respectively. Parish (1970) also demonstrated the presence of heat-stable IgG antibodies by passive cutaneous anaphylaxis on monkey skin. These results were confirmed by Biagini et al. (1985).

The sensitivity and reliability of the skin prick test has not been achieved in any *in vitro* test available. In enzyme immunoassays (Zachgo et al., 1985; Merget et al., 1988) and in the radioallergosorbent test (RAST) (Cromwell et al., 1979; Pepys et al., 1979) IgE antibodies specific to platinum chloride complexes were found. Although a good correlation with the results of prick tests was reported (Cromwell et al., 1979), their practical application for screening purposes was questioned because of the lack of specificity (Boggs, 1985; Jacobs, 1987; Merget et al., 1988). This was shown in a cross-sectional survey of platinum refinery workers (Merget et al., 1988). Higher total serum IgE and hexachloroplatinate-specific IgE levels in subjects with work-related symptoms were noted. However, not all the individuals allergic to platinum salt and some of the controls showed binding in RAST. Similar effects were seen with *in vitro* histamine release from basophils, which was relatively high in skin-test positive workers but even higher in the atopic control group. Histamine release with anti-IgE showed a similar pattern, indicating identical binding sites of hexachloroplatinate and anti-IgE on the surface of cutaneous mast cells and basophils.

Biagini et al. (1985) also found significantly higher mean total serum IgE levels in platinum refinery workers. However, good correlation between the RAST data and the skin test results was seen. Of the workers with positive skin tests, 95% (20/22) showed higher RAST binding than a control group, whereas only 8.5% (8/94) of those with negative skin tests demonstrated positive RAST results.

Since refinery workers are exposed to more than one platinum-group metal salt, the question of cross-reactivity was investigated by passive cutaneous anaphylaxis (PCA) tests. First results indicated that platinum ($Na_2[PtCl_6]$ and $(NH_4)_2[PtCl_6]$) and palladium ($Na_2[PdCl_6]$) appear to be equally effective as eliciting agents. . Five-fold concentrated sera from platinum refinery workers produced positive PCA results in monkeys (Biagini et al., 1982). No *in situ* reactions due to palladium salts were reported. There was only limited cross-reactivity between platinum and palladium salts in both skin test and RAST. Reactions to the platinum-group metals other than platinum were only seen in individuals sensitive to platinum salts (Murdoch et al., 1986; Murdoch & Pepys, 1987).

Instillation in the nose of concentrations of 10^{-3} to 10^{-9} g per ml has been used in the past as another method of detecting platinum salt sensitivity. A nasal reaction was considered positive if itching, sneezing, nasal obstruction or discharge occurred singly or in combination within 15 min of the challenge (Pepys et al., 1972).

Inhalation tests with a mixture of ammonium hexachloroplatinate and lactose dust gave immediate asthmatic reactions in sensitized individuals and in one case a late asthmatic reaction occurred (Pepys et al., 1972).

Merget et al. (1990) reported three cases of platinum refinery workers with negative skin tests who showed non-specific hyper-reactivity and a clearly positive immediate reaction in the bronchial provocation test.

8.2.5 Predisposing factors

Dally et al. (1980) conducted a retrospective cohort analysis in a group of 86 platinum workers entering a United Kingdom refinery in 1973-1974. It was found that significantly more atopics left employment, but this was apparently irrespective of the development of platinum salt allergy. The incidence of the disease did not differ significantly between the atopics (14/32 = 44%) and non-atopics (21/54 = 39%), although Burge et al. (1979) demonstrated that the atopics were sensitized more quickly. Thus, the increased leaving rate of atopics cannot be regarded as proof for the atopic status being a true predisposing factor, as suggested by Linnett (1987). It may, at most, be considered a trend.

Merget et al. (1988) examined 27 refinery workers and found no evidence to support tobacco smoking as a predisposing factor. However, Linnett (1987) found a significant association between smoking and the incidence of positive skin test results in life table studies of 134 refinery workers. Also, in a longitudinal cohort study on 91 platinum refinery workers (86 males, 5 females) in the United Kingdom who started work in 1973-1974 and were followed up until 1980, the risk of a positive skin test result was found to be 4-5 times higher in smokers than in non-smokers (Venables et al., 1989). Age, varying from 15-54 years in the cohort, was a definite confounding factor. After taking account of age, the risk of leaving refinery work was only 1.75 times greater in smokers than in non-smokers. The risk from atopy was not significant after taking smoking into consideration.

Brooks et al. (1990) further studied 107 current and 29 medically terminated workers, first described by Biagini (1985), using platinum skin testing and cold air challenge for evaluation of pulmonary hypersensitivity. Of these workers (74 current and 12 terminated workers), 63% underwent repeat platinum skin testing one year later. Among current workers, there was a conversion to positive platinum skin tests in five employees (with three of these conversions occurring in workers who had positive cold air challenge tests a year earlier). Thus, positive cold air challenge (airway hyperactivity) appears to have value for predicting conversion to positive skin test status with continued occupational exposure.

8.3 Side effects of cisplatin[a]

The therapeutic use of cisplatin is often complicated by the occurrence of side effects. Prominent among these are nephrotoxicity, severe nausea and vomiting, myelotoxicity (bone marrow suppression), and ototoxicity.

The most important toxic effect of cisplatin occurs in the kidney, eventually becoming irreversible during continued treatment. For instance, Lippman et al. (1973) found an approximately 50% reduction in renal function in each of 16 patients after treatment with total doses of 3.0-7.0 mg/kg body weight. Degeneration and necrosis of the proximal convoluted tubules, dilation of distal tubules, and glomerular abnormalities (elevation of the blood urea nitrogen and serum creatine levels, decreased

---

[a] See footnote in section 1.2.

Effects on Humans

creatine clearance) have been reported (Swierenga et al., 1987). A significant protection of renal function can be obtained by forced hydration, which flushes the drug through the kidney rapidly (Merrin, 1976). The simultaneous intravenous administration of mannitol can contribute to the prevention of cisplatin nephrotoxicity (Fillastre & Raguenez-Viotte, 1989).

Gastrointestinal toxicity consists mainly of nausea and vomiting lasting from 4 to 6 h and, occasionally in some sensitive patients, anorexia for up to one week (Hill et al., 1975).

Ototoxicity is another serious side-effect, consisting of tinnitus with or without clinical loss of hearing. Early in its course it is almost exclusively associated with high-range hearing loss in the 4000-8000 Hz range (Von Hoff et al., 1979).

Cisplatin can also cause peripheral neuropathy described as sensory, affecting primarily large fibres (Mollman et al., 1988).

Single cases of allergic reactions, angioneurotic oedema, rash, asthma (Von Hoff et al., 1979), cardiac arrest (Vogl et al., 1980), gingival discolouration (Ettinger & Freeman, 1979), and tetany due to hypocalcaemia and hypomagnesaemia (Hayes et al., 1979) have all been reported.

There are less toxic analogues, for example, *cis*-diammine-1,1-cyclobutanedicarboxylato platinum(II) (Carboplatin, JM8) and *cis*-dichloro-*trans*-dihydroxybisisopropylamine platinum(IV) (JM9). These cause less kidney damage, nausea, and vomiting. However, these analogues affect bone marrow and, in addition to the negative effects of cisplatin, may inhibit the formation of white cells, red cells, and blood platelets (Schacter & Carter, 1986; Bradford, 1988).

8.4 Carcinogenicity

No data are available to assess the carcinogenic risk of platinum or its salts to humans. With respect to cisplatin, IARC (1987b) considered the evidence for carcinogenicity for humans to be inadequate (see chapter 13).

9. EFFECTS ON OTHER ORGANISMS IN THE LABORATORY AND FIELD

9.1 Microorganisms

Simple complexes of platinum have bactericidal effects. In general, charged complexes in solutions, e.g., $(NH_4)_2[PtCl_6]$ above a concentration of 1 mg/litre, are lethal for bacteria. Neutral complexes are bactericidal only at considerably higher concentrations (> 38 µmol per litre) (Rosenberg et al., 1967; Shimazu & Rosenberg, 1973).

Rosenberg et al. (1965) reported the discovery of a unique property of some simple platinum-group metal complexes. When culture medium was subjected to an alternating current using platinum electrodes, bacterial cell division was inhibited. The spent medium itself was bactericidal. Detailed analysis revealed that the active agent was the cis isomer of $[PtCl_2(NH_3)_2]$, i.e. cisplatin. It was shown that such neutral platinum complexes, diluted in growth media, selectively inhibit cell division without reducing the cell growth of a variety of gram-positive and especially of gram-negative bacteria. As a result the bacterial rods are forced to form long filaments. This effect has been studied most intensively on *Escherichia coli* with cisplatin. In this case filamentation is reversed as soon as the bacterial filaments are transferred to a fresh medium free of the drug (Rosenberg et al., 1967).

Hoffmann (1988) studied the effects of cisplatin and $PtCl_4$ on the *in vitro* metabolism of the yeast *Saccharomyces cerevisiae*. Both compounds strongly inhibited DNA, RNA, and ribosome synthesis in the mmol/litre range. The IC_{50} (median inhibitory concentration) for the inhibition of DNA synthesis, for instance, was 0.42 mmol/litre (126 mg/litre) for cisplatin and 0.2 mmol/litre (67 mg per litre) for $PtCl_4$.

9.2 Aquatic organisms

9.2.1 Plants

Barnes & Talbert (1984) studied the influence of hexachloroplatinic acid (250, 500, and 750 µg/litre) on the growth of the green alga *Euglena gracilis* using a laboratory "microcosm". The growth recorded over 32 days was relatively slow, indicating that the experiments were conducted under low nutrient conditions, although these were not reported by the authors. For example, the doubling time of the control culture was about 9 days. Although no precise data were given, $H_2[PtCl_6]$ reduced growth rate and yield after 32 days in a dose-dependent manner.

After cisplatin was applied to the water hyacinth *Eichhornia crassipes* at 2.5 mg/litre, chlorosis was evident and the plants were stunted. At the 10-mg/litre level, some plant leaves were necrotic and chlorotic, and the roots were darkened and stunted. The most prominent symptom was the appearance of reddish-brown streaks on the leaves. These were particularly noticeable on young leaves and on the leaves of daughter plants (Farago & Parsons, 1985).

9.2.2 Animals

Biesinger & Christensen (1972), using Lake Superior water as a medium, studied the effects of various metals on the survival, growth, reproduction, and metabolism of the invertebrate *Daphnia magna*. Chronic (3-week) exposure to hexachloroplatinic acid, $H_2[PtCl_6]$, resulted in an LC_{50} value of 520 µg Pt/litre (range 437-619 µg per litre). Biochemical measurements and reproductivity were much more sensitive parameters than growth. A dose of 62 µg/litre caused a 12% reduction in weight gain, 13% reduction in total protein, and 20% decrease in glutamic-oxalacetic transaminase activity. At concentrations of 14 and 82 µg/litre, reproduction, measured as total number of young, was impaired by 16 and 50%, respectively.

Ferreira & Wolke (1979) investigated the effects of short-term exposure to tetrachloroplatinic acid, $H_2[PtCl_4]$, on the coho salmon *Oncorhynchus kisutch* at 8.5°C and a water hardness of about 56 mg $CaCO_3$/litre. In the static bioassay, 24-, 48-, and 96-h LC_{50} values of 15.5, 5.2, and 2.5 mg Pt/litre, respectively, were found. General swimming activity and opercular movement started to be affected at 0.3 mg/litre. Lesions in the gills and the olfactory organ were also noted at 0.3 mg/litre or more. Concentrations of 0.03 and 0.1 mg/litre had no effect.

9.3 Terrestrial organisms

A few studies have examined the effects of platinum on plants. All were conducted with soluble platinum chlorides.

Hamner (1942) investigated the effect of hexachloroplatinic acid, $H_2[PtCl_6].6H_2O$, on the growth of beans and tomato plants grown in sand culture. At concentrations of 3×10^{-5} to 15×10^{-5} mol/kg (5.9-29.3 mg/kg), growth was inhibited and the plants showed smaller leaf areas, higher osmotic pressure, and lower transpiration rates. They also resisted wilting longer than the controls and were less succulent.

Tso et al. (1973) reported that platinum increased the nicotine content of tobacco plants.

In a study by Pallas & Jones (1978) on the uptake of platinum by nine horticultural crops (see section 4.1), effects on growth were observed. Radish *(Raphanus sativus)*, cauliflower *(Brassica oleracea* cv. Snowball), snapbean *(Phaseolus vulgaris)*, sweet corn *(Zea mays)*, pea *(Pisum sativum)*, tomato *(Lycopersicon esculentum)*, bell pepper *(Capsicum annuum)*, broccoli *(Brassica oleracea* cv. Crusader), and turnips *(Brassica rapa)* were grown in hydroponic solution at 25/20 °C, 60/90% relative humidity, and 320/400 µl CO_2/litre air for 14/10 h photo- and nyctoperiods, respectively. When the seedlings reached an early maturity stage, such as flowering in the case of peas, snapbeans, cauliflower, tomato, and broccoli, root expansion in the case of turnip and radish, and considerable leafiness in the case of corn, platinum tetrachloride, $PtCl_4$, was added to fresh nutrient solution to give concentrations of 0.057, 0.57, and 5.7 mg Pt/litre. After a 7-day treatment, roots and tops were harvested and dried at 80 °C. As shown in Table 17, dry weights were significantly reduced in tomato, bell pepper, and turnip tops, and in radish roots at the highest platinum concentration. At this level, the buds and immature leaves of most species became chlorotic. In some of the species the low levels of $PtCl_4$ had a stimulatory effect on growth. In addition, transpiration was suppressed at the highest platinum concentration, probably due to increased stomatal resistance. Photosynthesis was also apparently reduced,

Table 17. Dry weight (g) of tops and roots after a 7-day treatment with platinum tetrachloride[a]

Pt levels (mg/litre)	Bean	Broccoli	Cauliflower	Corn	Pea	Pepper	Radish	Tomato	Turnip
Tops									
0	5.39	24.15	41.53	33.15	6.13	23.44	1.30	32.22	19.33
0.057	5.79	19.50	46.30	35.05	7.21	18.50	1.33	37.97	20.78
0.57	4.98	15.53	43.44	35.21	6.87	11.90	1.23	40.62	17.55
5.7	4.01	18.24	40.96	25.80	5.22	14.18	0.91	28.18	9.55
Roots									
0	3.11	5.45	6.09	9.96	2.86	4.82	2.17	5.96	4.80
0.057	2.80	4.77	6.56	10.96	2.60	4.82	2.24	7.64	5.51
0.57	2.70	4.02	5.68	10.73	2.81	3.32	2.17	6.68	4.91
5.7	2.31	5.36	6.57	8.36	2.20	4.90	1.19	6.00	3.82

[a] Adapted from: Pallas & Jones (1978)

consistent with the observed growth depression. On the other hand, the stimulation of transpiration and growth observed at the two lower concentrations, as compared to the control plants, explains the stimulated growth.

A stimulation of growth was also observed in seedlings of *Setaria verticillata* (L. P. Beauv) treated with a low level of platinum (0.5 mg Pt/litre) administered as potassium tetrachloroplatinate, $K_2[PtCl_4]$ (Farago & Parsons, 1986). This South African grass species was grown in nutrient solution, and after two weeks, the length of the longest roots had increased by 65%. At the higher concentration applied, i.e. 2.5 mg Pt/litre, phytotoxic effects were seen in the form of stunted root growth, i.e. root length about 75% compared to control, and chlorosis of the leaves. As platinum was shown to accumulate in the roots and, at the higher level, also in the shoots (see section 4.1), the potential use of this grass species, either for the colonisation of flotation tailings (a waste product from the concentration of precious metal ores) or for the removal of platinum from the tailings, was investigated. However, due to a substantial lack of essential macronutrients in the tailings, growth of *S. verticillata* was very poor. No platinum was detected in the plants.

10. EVALUATION OF HUMAN HEALTH RISKS AND EFFECTS ON THE ENVIRONMENT

10.1 Evaluation of human health risks

10.1.1 General population exposure

10.1.1.1 Exposure

There is lack of data on the actual exposure situation in countries where automobile exhaust gas catalysts have been introduced. Therefore, estimates of possible ambient air concentrations of platinum are based on emission data and dispersion models.

Loss of platinum from the pellet-type catalyst, which has never been used in Europe and is no longer used on new cars in the USA, was found to be up to approximately 2 µg per km travelled. Of the particles emitted, 80% had particle diameters greater than 125 µm. Since no determination of the particle size distribution was performed, the percentage of the respirable portion is not known. About 10% of the platinum emission was found to be water-soluble. In general, these data are based on single or only a few measurements and have not been validated.

Recent emission data from the new-generation monolith-type catalyst indicate that the emission of platinum is lower by a factor of 100-1000 than that of the pelleted catalyst. Emissions were on average between 2 and 39 ng per km travelled at simulated speeds of between 60 and 140 km/h. The mean aerodynamic diameter of the particles emitted was 4-9 µm. These data have been validated by repeated measurements using two monolith-type catalysts. However, other types of catalysts should be investigated to substantiate these emission data. In addition, because of the inadequate data base, the speciation is not known exactly, although there is an indication that the platinum emitted is in the metallic form or consists of surface-oxidized particles.

The striking difference between the emission pattern of the two catalyst types may be attributable to their basically different design.

The possible ambient air concentrations of platinum, estimated on the basis of these emission data and dispersion models, range between 0.005 and 9 ng/m^3 for the pellet-type catalyst and between 0.05 and 90 pg/m^3 for the monolith-type catalyst. These concentrations are lower by factors of 1×10^5 to 2×10^8 and 1×10^7 to 2×10^{10}, respectively, than the occupational exposure limit of 1 mg/m^3 established by some countries for platinum metal as total inhalable dust.

Assuming that 10% of the platinum emission from pelleted catalysts contains potentially allergenic soluble platinum compounds, the safety factor to the occupational exposure limit for soluble platinum salts (2 μg/m^3) would be 2×10^3 to 4×10^6. However, there is no evidence that the soluble fraction of the platinum emissions is allergenic.

0.1.1.2 Health effects

Since platinum is most probably not emitted in the form of halogenated soluble salts, the sensitization risk from car catalyst platinum is very low. There is no substantial evidence of any biological effects from automobile platinum emissions. There are also no data to substantiate the possibility that very finely dispersed metallic platinum could be biologically active upon inhalation.

10.1.2 Occupational groups

0.1.2.1 Exposure

Occupational exposure to platinum occurs in various workplaces including mining. However, only exposure to certain halogenated soluble salts through inhalation of dusts and skin contact is of toxicological relevance. These compounds are mainly encountered during platinum refining and catalyst manufacture.

There are only limited data to quantify workplace exposure. An occupational exposure limit of 2 μg/m^3 for soluble platinum salts has been adopted by several countries. There are again limited data suggesting that the exposure limit may sometimes be exceeded in practice.

Before the allergenic potential of soluble platinum salts was established, workplace concentrations exceeding the present occupational exposure limit by up to one order of magnitude were found. However, it should be noted that analytical accuracy was not very reliable.

Occupational exposure to the anti-tumour agents cis-platin and its analogues during manufacturing and use is of importance. However, a review and an evaluation of the health effects of these compounds are beyond the scope of this document as these substances are used primarily as therapeutic agents. In addition, their toxicological properties are exceptional compared to other platinum compounds.

10.1.2.2 Health effects

The acute toxicity of platinum salts for animals is low and depends on their solubility. Insoluble compounds such at $PtCl_2$ and PtO_2 have an extremely low acute toxicity and this would also be expected for metallic platinum.

By far the most significant health effect from exposure to soluble platinum salts is sensitization.

Some halogenated platinum salts are highly allergenic in humans. The compounds mainly responsible for platinum salt hypersensitivity (PSH) are hexachloroplatinic acid, $H_2[PtCl_6]$, ammonium hexachloroplatinate, $(NH_4)_2[PtCl_6]$, and potassium tetra- and hexachloroplatinate, $K_2[PtCl_4]$ and $K_2[PtCl_6]$. Except for one unsubstantiated case of alleged contact dermatitis in connection with a "platinum" ring, there is no evidence for sensitization from metallic platinum.

The mechanism of platinum salt allergy appears to be a type I (IgE mediated) response as established through *in vivo* and *in vitro* tests. There is evidence that platinum salts of low relative molecular mass act as haptens combining with serum proteins to form the complete antigen.

The signs and symptoms of allergic reactions due to platinum salt exposure include urticaria, itching of skin, eyes, and nose, watering of the eyes, sneezing, rhinorrhoea, coughing, wheezing, and dyspnoea. The latent period from the first contact with platinum salts to the occurrence of the first symptoms varies from a few weeks to several years. Once sensitivity is established, even minute amounts can elicit immediate and/or late onset of signs and symptoms. The symptoms persist during exposure and usually disappear on removal from exposure. However, if long-duration exposure occurs after sensitization, individuals may never become completely free of symptoms.

The diagnosis of platinum salt hypersensitivity is usually based on a history of work-related symptoms and a positive platinum skin prick test. The combination of these has been shown to be reasonably sensitive and specific for the diagnosis of platinum salt hypersensitivity. *In vitro* tests appear to be useful for epidemiological evaluation, but lack specificity for individual dignosis. Symptoms usually worsen as long as the workers remain in the contaminated environment.

There is good evidence for the association of smoking or pulmonary hyper-reactivity and sensitization. The evidence for atopy as a predisposing factor is equivocal. This may be due to bias from pre-employment screening.

Despite the occupational exposure limit of 2 $\mu g/m^3$, the prevalence of positive skin prick tests was found to be between 14 and 20% in workers exposed to levels of between < 0.05 and 0.1 μg Pt/m^3. Since these data were derived from area samples, short sharp exposures above this limit could also have been responsible for the sensitization observed. The present occupational exposure limit might not be sufficient to prevent platinum salt hypersensitization, although it is difficult to reach a firm conclusion because of the lack of adequate data. To minimize the risk, workplace exposure should be as low as practicable.

10.2 Evaluation of effects on the environment

Compared to that of other metals, the total production of platinum is low, amounting annually to approximately 100 tonnes. There are no data on platinum emissions during production. However, because of the high value of platinum, losses are assumed to be low. During the use of platinum-containing catalysts, platinum can escape into the environment in small amounts, depending on the type of catalyst. Of the stationary catalysts used in industry, only those used for ammonia oxidation emit a quantifiable amount of platinum. This is present in the nitric acid produced, which may be used for conversion to nitrate fertilizers. In the USA the annual loss of platinum is estimated to be around 200 kg. Since part of this amount is distributed fairly uniformly all over the country, a rise in the background level of platinum in soil would probably not be detected because of the very low likely concentration.

Platinum emission from automobile catalysts also contributes to a diffuse contamination of the environment. On the basis of the emission data derived from the new-generation monolithic-type catalysts, total platinum loss from mobile sources would be less than that from nitric acid production. For example, at an assumed average emission rate of 20 ng platinum per km travelled, 100 million cars equipped with catalytic converters would emit approximately 20 kg per year for an average kilometreage of 10 000 km per year and per car. This implies that the contamination of the environment with platinum is very low or negligible.

In comparison, the total loss of platinum from the older design pellet-type catalytic converter would have been higher by a factor of 100, i.e. 2000 kg per year, most of the platinum being emitted in the form of larger particles that would be deposited close to the roads. This would also explain platinum levels of up to 0.7 mg/kg dry weight found in roadside dust samples near major freeways in the USA.

There is limited evidence that most of the platinum emitted is in the metallic form, and thus will probably not be bioavailable in the soil. Biomethylation of soluble platinum(IV) compounds has been demonstrated in the presence of platinum(II). However, from these laboratory data produced under abiotic conditions, it is not possible to conclude that microorganisms in the environment are able to biomethylate platinum complexes.

Analysis of Lake Michigan sediments led to the conclusion that platinum has been deposited over the past 50 years at a constant rate, while lead concentrations have markedly increased. However, since the car catalyst was introduced in the USA only a few years before these measurements were performed, these data are insufficient for firm conclusions to be drawn.

No data on the effects of platinum compounds on environmental microorganisms are available. However, from the bactericidal activity of platinum complexes it can be assumed that these compounds could influence, at appropriate concentrations, microbial communities in the environment or, for example, in sewage treatment plants.

Aquatic and terrestrial plants are affected by platinum compounds at concentrations in the mg/litre or mg/kg range. Although there is a lack of definite data on platinum levels in the environment, it is probable that platinum and platinum compounds do not present a risk to naturally occurring organisms at the low concentrations expected to occur in the environment.

11. RECOMMENDATIONS FOR PROTECTION OF HUMAN HEALTH AND THE ENVIRONMENT

11.1 Pre-employment screening and medical evaluations

To screen workers at risk of developing platinum salt hypersensitivity (PSH), the following should be carried out for all employees potentially exposed to soluble platinum salts:

- a questionnaire with particular attention being paid to previous respiratory disease, allergy, smoking habits, and employment history;

- a complete medical examination, including lung function tests (spirometry, flow volume), tests of bronchial reactivity (cold air, methacholine, histamine, etc.), and an immunological profile including total serum IgE;

- a skin prick test for atopic status using a battery of antigens to include house dust mite, tree and grass pollen or other equivalent common aeroallergens;

- skin prick tests with freshly prepared, properly buffered saline solutions (e.g., 5% v/v glycerol/water containing 2.5 g NaCl/litre, 1.37 g $NaHCO_3$/litre, and 2 g phenol per litre) of $(NH_4)_2[PtCl_6]$, $Na_2[PtCl_4]$, and $Na_2[PtCl_6]$. Concentrations used for testing may vary from 10^{-9} to 10^{-3} g/ml depending on specific situations. All tests should be performed in duplicate and should include a positive and negative saline control.

11.2 Substitution with non-allergenic substances

An attempt should be made to substitute, whenever practicable, non-allergenic for allergenic platinum compounds during refining, manufacturing, and use.

11.3 Employment screening and medical evaluations

a) To detect sensitization during employment, skin prick tests should be performed on all potentially exposed people at least once a year. There is no convincing evidence that repeated platinum skin testing could cause sensitization. Quarterly testing intervals might be considered during the first two years of employment, as sensitization more often occurs during this period.

b) If medical symptoms or signs suggest the development of PSH, the worker should be removed from any risk of exposure as soon as possible. To detect functional changes in the respiratory tract, lung function assessment, as described in section 11.1, should also be performed at appropriate intervals.

11.4 Workplace hygiene

a) Since PSH can occur despite time-weighted average workplace concentrations being consistently below the ACGIH threshold limit value (TLV) of 2 $\mu g/m^3$, the most effective prevention is the improvement of control measures. This includes enclosed processing and optimal ventilation in order to reduce exposure to platinum salt aerosols and dusts to the lowest practicable limit.

b) It has been suggested that high but short-lived platinum concentrations resulting from spills or accidents are of importance with respect to sensitization. Since the correlation between the platinum exposure concentration and the development of sensitization is unknown, a recommendation for a reduction in the occupational exposure limit cannot be justified. However, it is recommended that the commonly used occupational exposure limit of 2 $\mu g/m^3$ be changed from an 8-h time-weighted average (TWA) to a ceiling value and that personal sampling devices be used in conjunction with area sampling to determine more correctly the true platinum exposure.

c) Engineering controls should always be in place to minimize exposure. However, in some circumstances the use of protective clothing, including specially designed airstream helmets, may be necessary.

d) Workers should be provided with clean overalls solely for use in the workplace, and showering facilities. Outdoor clothes should not be worn in the workplace.

12. FURTHER RESEARCH

a) As there appears to be a lack of information concerning the concentration-response relationship for the development of PSH in experimental animals, studies should be performed to investigate the effect of exposure concentration on sensitization and to define the thresholds for sensitization and elicitation.

b) The effect of predisposing factors such as pulmonary hyper-reactivity should be investigated in greater detail to determine their applicability for screening and identifying individuals at risk of developing PSH.

c) The use of provocation challenge with soluble platinum salts as an indicator of sensitization should be investigated to determine if it is a more sensitive indicator than skin prick tests.

d) The majority of human occupational studies regarding PSH were performed as cross-sectional studies at platinum refineries. Due to the inherent lack of sensitivity of this type of study with respect to past exposures and workers leaving employment because of disease, longitudinal studies should be performed to determine the true incidence of PSH in worker populations. In addition, human studies should be designed to study, for instance, exposure concentration effects on sensitization and determine thresholds for sensitization and elicitation.

e) The extent of occupational and environmental exposure to cisplatin is not known at the present time. It is recommended that studies be initiated to determine exposure during the manufacture and use of this compound.

f) Further measurements of the quantities and speciation of platinum emitted from automobile catalysts should be performed.

g) The toxic effects of finely divided metallic platinum on humans and animals have not been studied adequately. Adequate inhalation studies are initially required, and further tests may be necessary.

h) Quality control programmes should be initiated to ensure the accuracy and precision of sampling methods and analyses and to facilitate comparability.

i) Platinum-containing exhaust emissions from automobile catalysts most probably do not pose an adverse health effect for the general population. However, to be on the safe side, the possibility should be kept under review.

13. PREVIOUS EVALUATIONS BY INTERNATIONAL BODIES

The carcinogenicity of platinum and platinum compounds has not been evaluated by international bodies, except for cisplatin, which has not been covered in detail in this Environmental Health Criteria monograph (see also footnote in section 1.2).

The International Agency for Research on Cancer (IARC, 1987b) considered the evidence for carcinogenicity of cisplatin for animals to be sufficient, but that for humans inadequate. Cisplatin is classified in Group 2A, i.e. probably carcinogenic to humans.

REFERENCES

ACGIH (1980) Documentation of the threshold limit values, 4th ed., Cincinnati, Ohio, American Conference of Governmental Industrial Hygienists, pp. 343-344.

ACGIH (1990) Threshold limit values for chemical substances and physical agents and biological exposure indices for 1990-1991, Cincinnati, Ohio, American Conference of Governmental Industrial Hygienists, p. 31.

AGIORGITIS, G. & GUNDLACH, H. (1978) [Platinum contents of deep sea manganese nodules.] Naturwissenschaften, 65: 534 (in German).

AGNES, G., HILL, H.A.O., PRATT, J.M., RIDSDALE, S.C., KENNEDY, F.S., & WILLIAMS, R.J.P. (1971) Methyl transfer from methyl vitamin B_{12}. Biochim. Biophys. Acta, 252: 207-211.

AGOOS, A. (1986) Metal reclaimers find slim pickings. Chem. Week, 138: 21-23.

ALEXANDER, P.W., HOH, R., & SMYTHE, L.E. (1977a) Trace determination of platinum, I. A catalytic method using pulse polarography. Talanta, 24: 543-548.

ALEXANDER, P.W., HOH, R., & SMYTHE, L.E. (1977b) Trace determination of platinum, II. Analysis in ores by pulse polarography after fire-assay collection. Talanta, 24: 549-554.

ALT, F., JERONO, U., MESSERSCHMIDT, J., & TÖLG, G. (1988) The determination of platinum in biotic and environmental materials: I. µg/kg to g/kg-range. Mikrochim. Acta, 3: 299-304.

AMOSSE, J., FISCHER, W., ALLIBERT, M., & PIBOULE, M. (1986) Méthode de dosage d'ultra-traces de platine, palladium, rhodium et or dans les roches silicatées par spectrophotométrie d'absorption atomique électrothermique. Analusis, 14: 26-31.

ANDERSEN, K.S. (1979) Platinum(II) complexes generate frame-shift mutations in test strains of *Salmonella typhimurium*. Mutat. Res., 67: 209-214.

ANDREWS, P.A., WUNG, W.E., & HOWELL, S.B. (1984) A high-performance liquid chromatographic assay with improved selectivity for cisplatin and active platinum (II) complexes in plasma ultrafiltrate. Anal. Biochem., 43: 46-56.

ANEVA, Z., ARPADJAN, S., ALEXANDROV, S., & KOVATCHEVA, K. (1986) Synergetic extraction of platinum(IV) from dilute hydrochloric acid by isoamyl alcohol-methylisobutylketone mixture. Mikrochim. Acta, 1: 341-350.

ANON. (1990a) Nitric acid catalysts: the economic significance of composition, operation conditions and recovery techniques. Nitrogen, 183: 27-32.

ANON. (1990b) Growth slackens appreciably as 1980s end. Chem. Eng. News, 18 June: 36-43.

References

ANUSAVICE, K.J. (1985) Noble metal alloys for metal-ceramic restorations. Dent. Clin. North Am., 29: 789-803.

ARPADJAN, S., MITEWA, M., & BONTCHEV, P.R. (1987) Liquid-liquid extraction of metal ions by the 6-membered N-containing macrocycle hexacyclen. Talanta, 34: 953-956.

BANKOVSKY, Y.A., VIRCAVS, M.V., VEVERIS, O.E., PELNE, A.R., & VIRCAVA, D.K. (1987) Preconcentration of microamounts of elements in natural waters with 8-mercaptoquinoline and bis(8-quinolyl) disulphide for their atomic-absorption determination. Talanta, 34: 179-182.

BANNISTER, S.J., CHANG, Y., STERNSON, L.A., & REPTA, A.J. (1978) Atomic absorption spectrophotometry of free circulating platinum species in plasma derived from cis-dichlorodiammineplatinum(II). Clin. Chem., 24: 877-880.

BARLOW, S.M. & SULLIVAN, F.M. (1982) Reproductive hazards of industrial chemicals, London, Academic Press, pp. 431-436.

BARNES, G.D. & TALBERT, L.D. (1984) The effect of platinum on population and absorbance of *Euglena gracilis*, Klebs utilizing a method with atomic absorption and coulter counter analysis. J. Mississipi Acad. Sci., 29: 143-150.

BAUMGÄRTNER, M.E. & RAUB, Ch.J. (1988) The electrodeposition of platinum and platinum alloys. Platinum Met. Rev., 32: 188-197.

BELLIVEAU, J.F., MATOOK, G.M., O'LEARY, G.P., Jr, CUMMINGS, F.J., HILLSTROM, M., & CALABRESI, P. (1986) Microanalysis for platinum and magnesium in body fluids and brain tissue of mice treated with *cis*-platinum using graphite filament plasma emission spectroscopy. Anal. Lett., 19: 135-149.

BERNDT, H., SCHALDACH, G., & KLOCKENKÄMPER, R. (1987) Improvement of the detection power in electrothermal atomic absorption spectrometry by summation of signals. Determination of traces of metals in drinking water and urine. Anal. Chim. Acta, 200: 573-579.

BIAGINI, R.E., CLARK, J.C., GALLAGHER, J.S., BERNSTEIN, I.L., & MOORMAN, W.M. (1982) Passive transfer in the monkey of human immediate hypersensitivity to complex salts of platinum and palladium. Fed. Proc., 41: 827.

BIAGINI, R.E., MOORMAN, W.J., SMITH, R.J., LEWIS, T.R., & BERNSTEIN, I.L. (1983) Pulmonary hyperreactivity in cynomolgus monkeys (*Macaca fasicularis*) from nose-only inhalation exposure to disodium hexachloroplatinate, Na_2PtCl_6. Toxicol. appl. Pharmacol., 69: 377-384.

BIAGINI, R.E., BERNSTEIN, I.L., GALLAGHER, J.S., MOORMAN, W.J., BROOKS, S., & GANN, P.H. (1985) The diversity of reaginic immune responses to platinum and palladium metallic salts. J. Allergy clin. Immunol., 76: 794-802.

BIAGINI, R.E., MOORMAN, W.J., LEWIS, T.R., & BERNSTEIN, I.L. (1986) Ozone enhancement of platinum asthma in a primate model. Am. Rev. Respir. Dis., 134: 719-725.

BIESINGER, K.E. & CHRISTENSEN, G.M. (1972) Effects of various metals on survival, growth, reproduction, and metabolism of *Daphnia magna*. J. Fish Res. Board Can., 29: 1691-1700.

BODENNER, D.L., DEDON, P.C., KENG, P.C., KATZ, J.C., & BORCH, R.F. (1986) Selective protection against *cis*-diamminedichloroplatinum (II)-induced toxicity in kidney, gut, and bone marrow by diethyldithiocarbamate. Cancer Res., 46: 2751-2755.

BOGGS, P.B. (1985) Platinum allergy. Cutis, 35: 318-320.

BOLM-AUDORFF, U., BIENFAIT, H.-G., BURKHARD, J., BURY, A.-H., MERGET, R., PRESSEL, G., & SCHULTZE-WERNINGHAUS, G. (1988) [On the frequency of respiratory allergies in a platinum processing factory.] In: Baumgartner, E., Brenner, W., Dierich, M.P., & Rutenfranz, J., ed. [Industrial change - Occupational medicine facing new questions. Report of the 28th Annual Meeting of the German Society for Occupational Medicine, Innsbruck, Austria, 4-7 May 1988] Stuttgart, Gentner Verlag, pp. 411-416 (in German).

BORCH, R.F., MARKOVITZ, J.H., & PLEASANTS, M.E. (1979) A new method for the HPLC analysis of Pt(II) in urine. Anal. Lett., 12: 917-926.

BOUMANS, P.W.J.M. & VRAKKING, J.J.A.M. (1987) Detection of about 350 prominent lines of 65 elements observed in 50 and 27 MHz inductively coupled plasmas (ICP): effects of source characteristics, noise and spectral bandwidth - "Standard" values for the 50 MHz ICP. Spectrochim. Acta, 42B: 553-579.

BOWEN, H.J.M. (1979) Environmental chemistry of the elements, London, Academic Press, p. 37.

BRADFORD, C.W. (1988) Platinum. In: Seiler, H.G. & Sigel, H., ed. Handbook of toxicity of inorganic compounds, New York, Basel, Marcel Dekker, pp. 533-539.

BRAJTER, K. & KOZICKA, U. (1979) Extractive spectrophotometric determination of platinum, rhodium and iridium. Talanta, 26: 417-419.

BROOKS, S.M., BAKER, D.B., GANN, P.H., JARABEK, A.M., HERTZBERG, V., GALLAGHER, J., BIAGINI, R.E., & BERNSTEIN, I.L. (1990) Cold air challenge and platinum skin reactivity in platinum refinery workers. Bronchial reactivity precedes skin prick response. Chest, 79: 1401-1407.

BROWN, A.A. & LEE, M. (1986) Peak profile and appearance times using totally pyrolytic cuvettes in graphite furnace atomic absorption spectrometry. Anal. Chem., 323: 697-702.

BRUBAKER, P.E., MORAN, J.P., BRIDBORD, K., & HUETER, F.G. (1975) Noble metals: a toxicological appraisal of potential new environmental contaminants. Environ. Health Perspect., 10: 39-56.

BURGE, P.S., HARRIES, M.G., O'BRIEN, I.M., & PEPYS, J. (1979) Evidence for specific hypersensitivity in occupational asthma due to small molecular weight chemicals and an organic (locust) allergen. In: Pepys, J. & Edwards, A.M., ed. The mast cell - its role in health and disease, Tunbridge Wells, Kent, United Kingdom, Pitman Medical Publishing Co., Ltd, pp. 301-308.

References

BUTTERMAN, W.C. (1975) Platinum-group metals. Minerals Yearbook 1973 Metals, Minerals, and Fuels, Washington, DC, US Department of the Interior, Bureau of Mines, pp. 1037-1049.

CAMPBELL, K.I., GEORGE, E.L., HALL, L.L., & STARA, J.F. (1975) Dermal irritancy of metal compounds. Arch. environ. Health, 30: 168-170.

CHANG, C.C. & TATUM, H.J. (1975) Effect of intrauterine copper wire on resorption of foetuses in rats. Contraception, 11: 79-84.

CHANG, C.C., TATUM, H.J., & KINCL, F.A. (1970) The effect of intrauterine copper and other metals on implantation in rats and hamsters. Fertil. Steril. 21(3): 274-278.

CHEY, R.D., LEE, K.Y, ASBURY, R., & CHEY, W.Y. (1988) Effect of cisplatin on myoelectric activity of the stomach and small intestine in dogs. Dig. Dis. Sci., 33: 338-344.

CLEARE, M.J. (1977a) Some aspects of platinum complex chemistry and their relation to anti-tumor activity. J. clin. Hematol. Oncol., 7: 1-21.

CLEARE, M.J. (1977b) Immunological studies on platinum complexes and their possible relevance to autocatalysts. Proceedings of the International Automotive Engineering Congress and Exposition, Cobo Hall, Detroit, 28 February - 4 March 24, 1977, Detroit, Michigan Society of Automotive Engineers, 12 pp (SAE Report No. 770061).

CLEARE, M.J., HUGHES, E.G., JACOBY, B., & PEPYS, J. (1976) Immediate (type I) allergic responses to platinum compounds. Clin. Allergy, 6: 183-195.

COOMBES, R.J. & CHOW, A. (1979) A comparison of methods for the determination of platinum in ores. Talanta, 26: 991-998.

CROMWELL, O., PEPYS, J., PARISH, W.E, & HUGHES, E.G. (1979) Specific IgE antibodies to platinum salts in sensitized workers. Clin. Allergy, 9: 109-117.

CVITKOVIC, E., SPAULDING, J., BETHUNE, V., MARTIN, J., & WHITMORE, W.F. (1977) Improvement of cis-dichlorodiammineplatinum (NSC 119875): therapeutic index in an animal model. Cancer, 39: 1357-1361.

D'AGOSTINO, R.B., LOWN, B.A., MORGANTI, J.B., CHAPIN, E., & MASSARO, E.J. (1984) Effects on the development of offspring of female mice exposed to platinum sulfate or sodium hexachloroplatinate during pregnancy or lactation. J. Toxicol. environ. Health, 13: 879-891.

DALLY, M.B., HUNTER, J.V., HUGHES, E.G., STEWART, M., & NEWMAN TAYLOR, A.J. (1980) Hypersensitivity to platinum salts: a population study. Am. Rev. respir. Dis., 121(4/Part 2): 230.

DE CONTI, R.C., TOFTNESS, B.R., LANGE, R.C., & CREASEY, W.A. (1973) Clinical and pharmacological studies with cis-diammine-dichloroplatinum (II). Cancer Res., 33: 1310-1315.

DEGUSSA (1988a) Ammonium-tetrachloroplatinate(II) - Acute toxicity. Testing the primary irritancy after single application to the skin of the rabbit (patch test), Hanau, Germany, Degussa AG (Unpublished report No. 863897).

DEGUSSA (1988b) Ammonium-tetrachloroplatinate(II) - Acute toxicity. Testing the primary irritancy after single application to the eye of the rabbit, Hanau, Germany, Degussa AG (Unpublished report No. 863908).

DEGUSSA (1988c) Trans-diammine-dichloroplatinum(II) - Acute toxicity. Testing the primary irritancy after single application to the skin of the rabbit, Hanau, Germany, Degussa AG (unpublished report No. 864011).

DEGUSSA (1988d) Trans-diammine-dichloroplatinum(II) - Acute toxicity. Testing the primary irritancy after single application to the eye of the rabbit, Hanau, Germany, Degussa AG (Unpublished report No. 864022).

DEGUSSA (1989a) Ammoniumtetrachloroplatinate(II) - Acute toxicity after single oral administration in rats, Hanau, Germany, Degussa AG (Unpublished report No. 863886).

DEGUSSA (1989b) Dinitrodiammine-platinum(II) - Acute toxicity after single oral administration in rats, Hanau, Germany, Degussa AG (Unpublished report No. 863910).

DEGUSSA (1989c) Trans-diammine-dichloroplatinum(II) - Acute toxicity after single oral administration in rats, Hanau, Germany, Degussa AG (Unpublished report No. 864000).

DEGUSSA (1989d) Dinitrodiammine-platinum(II) - Acute toxicity. Testing the primary irritancy after single application to the skin of the rabbit (patch test), Hanau, Germany, Degussa AG (Unpublished report No. 863921).

DEGUSSA (1989e) Dinitrodiammine-platinum(II) - Acute toxicity. Testing the primary irritancy after single application to the eye of the rabbit), Hanau, Germany, Degussa AG (Unpublished report No. 863932).

DISSANAYAKE, C.B. (1983) Metal-organic interactions in environmental pollution. Int. J. environ. Stud., 22: 25-42.

DISSANAYAKE, C.B., KRITSOTAKIS, K., & TOBSCHALL, H.J. (1984) The abundance of Au, Pt, Pd, and the mode of heavy metal fixation in highly polluted sediments from the Rhine River near Mainz, West Germany. Int. J. environ. Stud., 22: 109-119.

DONALDSON, P.E.K. (1987) The role of platinum metals in neurological prostheses. Platinum Met. Rev., 31: 2-7.

DUFFIELD, F.V.P., YOAKUM, A., BUMGARNER, J., & MORAN, J. (1976) Determination of human body burden baseline data of platinum through autopsy tissue analysis. Environ. Health Perspect., 15: 131-134.

References

EBINA, T., SUZUKI, H., & YOTSUYANAGI, T. (1983) [Spectrophotometric determination of palladium(II) and platinum(II) with maleonitriledithiol by reversed phase ion-pair chromatography.] Bunseki Kagaku, 32: 575-580 (in Japanese).

ELLER, P.M., ed. (1984a) Elements (ICP). In: NIOSH manual of analytical methods, 3rd ed., Cincinnati, Ohio, National Institute for Occupational Safety and Health, Vol. 1, pp. 7300-1-7300-5.

ELLER, P.M., ed. (1984b) Elements in blood or tissue. In: NIOSH manual of analytical methods, 3rd ed., Cincinnati, Ohio, National Institute for Occupational Safety and Health, Vol.2, pp. 8005-1-8005-5.

ELLER, P.M., ed. (1984c) Metals in urine. In: NIOSH manual of analytical methods, 3rd ed., Cincinnati, Ohio, National Institute for Occupational Safety and Health, Vol. 2, pp. 8310-1-8310-6.

ETTINGER, L.J. & FREEMAN, A.I. (1979) The gingival platinum line. A new finding following cis-dichlorodiammine platinum(II) treatment. Cancer, 44: 1882-1884.

FANCHIANG, Y.-T. (1985) Reactions of alkylcobalamins with platinum complexes. Coord. Chem. Rev., 68: 131-144.

FANCHIANG, Y-T., RIDLEY, W.P., & WOOD, J.M. (1979) Methylation of platinum complexes by methylcobalamin. J. Am. Chem. Soc., 101: 1442-1447.

FARAGO, M.E. & PARSONS, P.J. (1982) Determination of platinum, palladium and rhodium by atomic-absorption spectroscopy with electrothermal atomisation. Analyst, 107: 1218-1228.

FARAGO, M.E. & PARSONS, P.J. (1985) Effects of platinum metals on plants. Trace Subst. environ. Health, 19: 397-407.

FARAGO, M.E. & PARSONS, P.J. (1986) The effect of platinum, applied as potassium tetrachloroplatinate, on *Setaria verticillata* (L) P. Beauv. and its growth on flotation tailings. Environ. Technol. Lett., 7: 147-154.

FERREIRA, P.F. & WOLKE, R.E. (1979) Acute toxicity of platinum to coho salmon (*Oncorhynchus kisutch*). Mar. Pollut. Bull., 10: 79-83.

FILLASTRE, J.P. & RAGUENEZ-VIOTTE, G. (1989) Cisplatin nephrotoxicity. Toxicol. Lett., 46: 163-175.

FOTHERGILL, S.J.R., WITHERS, D.F., & CLEMENTS, F.S. (1945) Determination of traces of platinum and palladium in the atmosphere of a platinum refinery. Br. J. ind. Med., 2: 99-101.

FOX, R.L. (1984) Enhancement errors in the determination of platinum group metals in alumina-based matrices by direct-current plasma emission spectrometry. Appl. Spectrosc., 38: 644-647.

FREEDMAN, S.O. & KRUPEY, J. (1968) Respiratory allergy caused by platinum salts. J. Allergy, 42: 233-237.

FRYER, B.J. & KERRICH, R. (1978) Determination of precious metals at ppb levels in rocks by a combined wet chemical and flameless atomic absorption method. At. Absorpt. Newsl., 17: 4-6.

FUCHS, W.A. & ROSE, A.W. (1974) The geochemical behavior of platinum and palladium in the weathering cycle in the Stillwater Complex, Montana. Econ. Geol., 69: 332-346.

GALLAGHER, J.S., BAKER, D., GANN, P.H., JARABEK, A.M., BROOKS, S.M., & BERNSTEIN, I.L. (1982) A cross sectional investigation of workers exposed to platinum salts. J. Allergy clin. Immunol., 69: 134.

GECKELER, K.E., BAYER, E., SPIVAKOV, B.Y., SHKINEV, V.M., & VOROB'EVA, G.A. (1986) Liquid-phase polymer-based retention; a new method for separation and preconcentration of elements. Anal. Chim. Acta, 189: 285-292.

GLADKOVA, E.V., ODINTSOVA, F.P., VOLKOVA, I.D., & VINOGRADOVA, V.K. (1974) [Health status of workers engaged in the production of the platinum catalyst.] Gig. Tr. prof.Zabol., 3: 10-13 (in Russian).

GOLDBERG, E.D., HODGE, V.F., GRIFFIN, J.J., KOIDE, M., & EDGINGTON, D.N. (1981) The impact of fossil fuel combustion on the sediments of Lake Michigan. Environ. Sci. Technol., 15: 466-471.

GOLDBERG, E.D., HODGE, V., KAY, P., STALLARD, M., & KOIDE, M. (1986) Some comparative marine chemistries of platinum and iridium. Appl. Geochem., 1: 227-232.

GOLDSCHMIDT, V.M. (1954) Geochemistry, London, Oxford University Press, 730 pp.

GREGOIRE, D.C. (1988) Determination of platinum, palladium, ruthenium and iridium geological materials by inductively coupled plasma mass spectrometry with sample introduction by electrothermal vaporization. J. anal. at. Spectrom., 3: 309-314.

GROOTVELD, M.C. (1985) Studies on the reaction mechanisms of metal complexes of biological and immunochemical interest, London, University of London, Department of Chemistry, 341 pp (Thesis).

GROTE, M. & KETTRUP, A. (1987) Ion-exchange resins containing S-bonded dithizone and dehydrodithizone as functional groups. Part 3. Determination of gold, platinum and palladium in geological samples by means of a dehydrodithizone resin and plasma emission spectrometry. Anal. Chim. Acta, 201: 95-107.

GSF (1990) [Noble metal emissions - interim report; Research Programme of the Federal Ministry of Research and Technology], Munich, Association for Radiation and Environmental Research (GSF), 45 pp. (in German).

HALBACH, P., PUTEANUS, D., & MANHEIM, F.T. (1984) Platinum concentrations in ferromanganese seamount crusts from the central Pacific. Naturwissenschaften, 71: 577-579.

References

HAMILTON, E.L. & MINSKI, M.J. (1972/1973) Abundance of the chemical elements in man's diet and possible relations with environmental factors. Sci. total Environ., 1: 375-394.

HAMNER, C.L. (1942) Effects of platinum chloride on bean and tomato. Bot. Gaz., 104: 161-166.

HARDER, H.C. & ROSENBERG, B. (1970) Inhibitory effects of anti-tumor platinum compounds on DNA, RNA and protein syntheses in mammalian cells in vitro. Int. J. Cancer, 6: 207-216.

HARDMAN, R.S. & WRIGHT, C.H. (1896) A case of poisoning by chloroplatinite of potassium. Br. med. J., 1: 529.

HAYES, F.A., GREEN, A.A., SENZER, N., & PRATT, C.B. (1979) Tetany: a complication of cis-dichlorodiammine-platinum(II) therapy. Cancer Treat. Rep., 63: 547-548.

HEALTH AND SAFETY EXECUTIVE (1983) The medical monitoring of workers exposed to platinum 'salts', London, Her Majesty's Stationery Office, 2 pp (MS/22).

HEALTH and SAFETY EXECUTIVE (1985) Platinum metal and soluble inorganic compounds of platinum in air. Laboratory method using carbon furnace atomic absorption spectrometry, London, Her Majesty's Stationery Office, 4 pp (MDHS 46).

HEALTH AND SAFETY EXECUTIVE (1990) Occupational exposure limits 1990. Guidance note EH 40190, London, Her Majesty's Stationery Office.

HEBERT, R. (1966) Affections provoquées par les composés du platine. Arch. Mal. prof. Méd. Trav. Sécur. soc., 27: 877-886.

HILL, J.M., LOEB, E., MACLELLAN, A., HILL, N.O., KHAN, A., & KING, J.J. (1975) Clinical studies of platinum coordination compounds in the treatment of various malignant diseases. Cancer Chemother. Rep., 59(Part 1): 647-659.

HILL, R.F. & MAYER, W.J. (1977) Radiometric determination of platinum and palladium attrition from automotive catalysts. IEEE Trans. nucl. Sci., NS-24: 2549-2554.

HODGE, V.F. & STALLARD, M.O. (1986) Platinum and palladium in roadside dust. Environ. Sci. Technol., 20: 1058-1060.

HODGE, V.F., STALLARD, M., KOIDE, M., & GOLDBERG, E.D. (1985) Platinum and the platinum anomaly in the marine environment. Earth planet. Sci. Lett., 72: 158-162.

HODGE, V.F., STALLARD, M., KOIDE, M., & GOLDBERG, E.D. (1986) Determination of platinum and iridium in marine waters, sediments, and organisms. Anal. Chem., 58: 616-620.

HOESCHELE, J.D. & VAN CAMP, L. (1972) Whole-body counting and the distribution of cis^{195m}[Pt(NH$_3$)$_2$Cl$_2$] in the major organs of Swiss white mice. In: Hejzlar, M., ed. Advances in antimicrobial and antineoplastic chemotherapy, Baltimore, University Park Press, Vol. 2, pp. 241-242.

HOEVE, L.J., CONIJIN, E.A.J.G., MERTENS ZUR BORG, I.R.A.M., & RODENBURG, M. (1987) Development of cis-platinum-induced hearing loss in guinea pigs. Arch. Otorhinolaryngol., 244: 265-268.

HOFFMANN, R.L. (1988) The effect of cisplatin and platinum (IV) chloride on cell growth, RNA, protein, ribosome and DNA synthesis in yeast. Toxicol. environ. Chem., 17: 139-151.

HOFMEISTER, F. (1882) [On the physiologic effects of the platinum bases.] Naunyn-Schmiedeberg's Arch. exp. Pathol. Pharmakol., 16: 393-439 (in German).

HOLBROOK, D.J., Jr (1976a) Assessment of toxicity of automotive metallic emissions, Vol. I: Assessment of fuel additives emission toxicity via selected assays of nucleic and protein synthesis, Research Triangle Park, North Carolina, US Environmental Protection Agency, Office of Research and Development, Health Effects Research Laboratories, 67 pp (EPA/600/1-76/010a).

HOLBROOK, D.J., Jr (1976b) Assessment of toxicity of automotive metallic emissions, Vol. II: Relative toxicities of automotive metallic emissions against lead compounds using biochemical parameters, Research Triangle Park, North Carolina, US Environmental Protection Agency, Office of Research and Development, Health Effects Research Laboratories, 69 pp. (EPA/600/1-76/010b).

HOLBROOK, D.J., Jr (1977) Content of platinum and palladium in rat tissue: correlation of tissue concentration of platinum and palladium with biochemical effects, Research Triangle Park, North Carolina, US Environmental Protection Agency, Office of Research and Development, Health Effects Research Laboratory, 21 pp. (EPA/600/1-77/051).

HOLBROOK, D.J., Jr, WASHINGTON, M.E., LEAKE, H.B., & BRUBAKER, P.E. (1975) Studies on the evaluation of the toxicity of various salts of lead, manganese, platinum, and palladium. Environ. Health Perspect., 10: 95-101.

HOPPE, R. (1965) [Chemistry of the noble gases.] Fortschr. chem. Forsch., 5: 213-346 (in German).

HOPPSTOCK, K., ALT, F., CAMMANN, K., & WEBER, G. (1989) Determination of platinum in biotic and environmental materials. Part II: A sensitive voltammetric method. Fresenius Z. anal. Chem., 335: 813-816.

HOWLE, J.A. & GALE, G.R. (1970) Cis-dichlorodiammineplatinum (II) persistent and selective inhibition of deoxyribonucleic acid synthesis *in vivo*. Biochem. Pharmacol., 19: 2757-2762.

HUGHES, E.G. (1980) Medical surveillance of platinum refinery workers. J. Soc. Occup. Med., 30: 27-30.

HUNTER, D., MILTON, R., & PERRY, K.M.A. (1945) Asthma caused by the complex salts of platinum. Br. J. ind. Med., 2: 92-98.

IARC (1987a) Cisplatin. In: Genetic and related effects: an updating of selected IARC Monographs from Volumes 1 to 42, Lyon, International Agency for Research on Cancer, pp. 178-181 (IARC Monographs on the Evaluation of Carcinogenic Risks to Humans, Suppl. 6).

IARC (1987b) Cisplatin (Group 2A). In: Overall evaluations of carcinogenicity: an updating of IARC Monographs Volumes 1 to 42, Lyon, International Agency for Research on Cancer, pp. 170-171 (IARC Monographs on the Evaluation of Carcinogenic Risks to Humans, Suppl. 7).

IMURA, N. (1986) The role of micronutrient selenium in the manifestation of toxicity of heavy metals. Toxicol. Lett., 31: 11.

INGALLS, M.N. & GARBE, R.J. (1982) Ambient pollutant concentrations from mobile sources in microscale situations, Warrendale, Pennsylvania, Society of Automotive Engineers, Inc., 16 pp (SAE Technical Paper Series. No. 820787).

ITO, E. & KIDANI, Y. (1982) Determination of platinum in the environmental samples by graphite furnace atomic absorption spectrophotometry. Bunseki Kagaku, 31: 381-388.

JACOBS, L. (1987) Platinum salt sensitivity. Nurs. RSA Verpleging (Republic of South Africa), 2: 34-37.

JOHNSON, D.E., TILLERY, J.B., & PREVOST, R.J. (1975) Levels of platinum, palladium, and lead in populations of Southern California. Environ. Health Perspect., 12: 27-33.

JOHNSON, D.E., PREVOST, R.J., TILLERY, J.B., CAMAN, D.E., & HOSENFELD, J.M. (1976) Baseline levels of platinum and palladium in human tissue, San Antonio, Texas, Southwest Research Institute, 252 pp. (EPA/600/1-76/019).

JOHNSON, N.P., HOESCHELE, J.D., RAHN, R.O., O'NEIL, P.J., & HSIE, A.W. (1980) Mutagenicity, cytotoxicity and DNA binding of platinum(II)-chloroammines in chinese hamster ovary cells. Cancer Res., 40: 1463-1468.

JOHNSON MATTHEY (1976a) Acute oral toxicity study CB10: Bis(acetylacetonate) platinum(II), Reading, United Kingdom, Johnson Matthey Research Centre (Unpublished report No. 8/7610)

JOHNSON MATTHEY (1976b) Skin irritation study, CB10: Bis(acetylacetonate) platinum(II), Reading, United Kingdom, Johnson Matthey Research Centre (Unpublished report No. 8/7610).

JOHNSON MATTHEY (1976c) Eye irritation study CB10: Bis(acetylacetonate) platinum(II), Reading, United Kingdom, Johnson Matthey Research Centre (Unpublished report No.8/7610).

JOHNSON MATTHEY (1977a) Acute oral toxicity study, CB20: Potassium tetracyanoplatinate(II), Reading, United Kingdom, Johnson Matthey Research Centre (Unpublished report No. 86/7709).

JOHNSON MATTHEY (1977b) Acute oral toxicity study, CB18: Tetrammine platinous chloride, Reading, United Kingdom, Johnson Matthey Research Centre (Unpublished report No. 248/7702).

JOHNSON MATTHEY (1977c) Acute oral toxicity study, CB18: Cis[PtCl$_2$(NH$_3$)$_2$], Reading, United Kingdom, Johnson Matthey Research Centre (Unpublished report No. 216/7707).

JOHNSON MATTHEY (1977d) Primary skin irritation study, CB20: Potassium tetracyanoplatinate(II), Reading, United Kingdom, Johnson Matthey Research Centre (Unpublished report No. 88/7709).

JOHNSON MATTHEY (1977e) Skin irritation study, CB13: Tetrammine platinous chloride, Reading, United Kingdom, Johnson Matthey Research Centre (Unpublished report No. 113/7701).

JOHNSON MATTHEY (1977f) Eye irritation study, CB13: Tetrammine platinous chloride, Reading, United Kingdom, Johnson Matthey Research Centre (Unpublished report No. 252/7702).

JOHNSON MATTHEY (1977g) Primary skin irritation study, CB18: Cis[PtCl$_2$(NH$_3$)$_2$], Reading, United Kingdom, Johnson Matthey Research Centre (Unpublished report No. 217/7707).

JOHNSON MATTHEY (1977h) Eye irritation/toxicity study, sample CB18: Cis[PtCl$_2$(NH$_3$)$_2$], Reading, United Kingdom, Johnson Matthey Research Centre (Unpublished report No. 337/7802).

JOHNSON MATTHEY (1978a) Acute oral toxicity study in the rat, CB14: Ammonium hexachloroplatinate, batch No. 57662, Reading, United Kingdom, Johnson Matthey Research Centre (Unpublished report No. 106/7808).

JOHNSON MATTHEY (1978b) Acute oral toxicity study, CB24: Sodium chloroplatinate, Reading, United Kingdom, Johnson Matthey Research Centre (Unpublished report No. 359/7711).

JOHNSON MATTHEY (1978c) Acute oral toxicity study in the rat, CB70: Sodium hexachloroplatinate, Batch No. 031074, Reading, United Kingdom, Johnson Matthey Research Centre (Unpublished report No. 42/7809).

JOHNSON MATTHEY (1978d) Primary skin irritation study, sample CB14: Ammonium hexachloroplatinate, batch No. 57662, Reading, United Kingdom, Johnson Matthey Research Centre (Unpublished report No. 109/7808).

JOHNSON MATTHEY (1978e) Primary skin irritation study, sample CB24: Sodium chloroplatinate, batch No. 031003, Reading, United Kingdom, Johnson Matthey Research Centre (Unpublished report No. 245/7802).

References

JOHNSON MATTHEY (1978f) Eye irritation study, sample CB24: Sodium chloroplatinate, batch No. 031003, Reading, United Kingdom, Johnson Matthey Research Centre (Unpublished report No. 240/7802).

JOHNSON MATTHEY (1978g) Primary skin irritation study U.S. Federal Register 1973, sample CB70: Sodium hexahydroxyplatinate, batch No. 031074, Reading, United Kingdom, Johnson Matthey Research Centre (Unpublished report No. 43/7809).

JOHNSON MATTHEY (1978h) Eye irritation study, sample CB20: Potassium tetracyanoplatinate(II), Reading, United Kingdom, Johnson Matthey Research Centre (Unpublished report No. 318/7802).

JOHNSON MATTHEY (1981a) An assessment of the acute oral toxicity of potassium tetracyanoplatinate(II)-MS 332 in the rat, experiment No. 24/8112, Reading, United Kingdom, Johnson Matthey Research Centre (Unpublished report).

JOHNSON MATTHEY (1981b) An assessment of the acute oral toxicity of potassium tetracyanoplatinate(II)-MS 332 in the rat, experiment No. 216/8111, Reading, United Kingdom, Johnson Matthey Research Centre (Unpublished report).

JOHNSON MATTHEY (1981c) OECD skin irritation test: determination of the degree of primary cutaneous irritation caused by potassium tetracyanoplatinate(II)-MS 332 in the rabbit, experiment No. 631/8111, Reading, United Kingdom, Johnson Matthey Research Centre (Unpublished report).

JOHNSON MATTHEY (1981d) OECD eye irritation test: determination of the degree of ocular irritation caused by potassium tetrachloroplatinate(II)-MS 332 in the rabbit, experiment No. 278/8112, Reading, United Kingdom, Johnson Matthey Research Centre (Unpublished report).

JOHNSON MATTHEY (1990) Platinum 1990, London, Johnson Matthey Public Limited Company, 64 pp.

JONES, A.H. (1976) Determination of platinum and palladium in blood and urine by flameless atomic absorption spectrophotometry. Anal. Chem., **48**: 1472-1474.

JONES, E.A., WARSHAWSKY, A., DIXON, K., NICOLAS, D.J., & STEELE, T.W. (1977) The group extraction of noble metals with s-(1-Decyl)-N,N'-diphenylisothiouronium bromide and their determination in the organic extract by atomic-absorption spectrometry. Anal. Chim. Acta, **94**: 257-268.

JUNG, H.J. & BECKER, E.R. (1987) Emission control for gas turbines. Platinum-rhodium catalysts for carbon monoxide and hydrocarbon removal. Platinum Met. Rev., **31**: 162-170.

KAHN, N. & VAN LOON, J.V. (1978) Direct atomic absorption spectrophotometric analysis of anion complexes of platinum and gold after their concentration and separation from aqueous solutions by anion exchange chromatography. Anal. Lett., **12**: 991-1000.

KANEMATSU, N., HARA, M., & KADA, T. (1980) Rec assay and mutagenicity studies on metal compounds. Mutat. Res., **77**: 109-116.

KARASEK, S.R. & KARASEK, M. (1911) The use of platinum paper. Report of (Illinois) Commission on Occupational Diseases to His Excellency Governor Charles S. Deneen, January 1911, Chicago, Warner Printing Company, p. 97

KENAWY, I.M., KHALIFA, M.E., & EL-DEFRAWY, M.M. (1987) Preconcentration and determination (AAS) of trace Ag(I), Au(III), Pd(II) and Pt(IV) using a cellulose ion-exchanger (Hyphan). Analusis, 15: 314-317.

KESSERÜ, E. & LEON, F., (1974) Effect of different solid metals and metallic pairs on human sperm motility. Int. J. Fertil., 19: 81-84.

KHAN, A., HILL, J.M., GRATER, W., LOEB, E., MACLELLAN, A., & HILL, N. (1975) Atopic hypersensitivity to cis-dichlorodiammineplatinum(II) and other platinum complexes. Cancer Res., 35: 2766-2770.

KNAPP, G. (1984) [Routes to powerful methods of elemental trace analysis in environmental samples.] Fresenius Z. anal. Chem., 317: 213-219 (in German).

KNAPP, G. (1985) Sample preparation techniques - An important part in trace element analysis for environmental research and monitoring. Int. J. environ. anal. Chem., 22: 71-83.

KOBERSTEIN, E. (1984) [Catalysts for purifying automotive exhaust gas.] Chem. Zeit, 18: 37-45 (in German).

KOCIBA, R.J. & SLEIGHT, S.D. (1971) Acute toxicologic and pathologic effects of cis-diamminedichloroplatinum (NSC-119875) in the male rat. Cancer Chemother. Rep., 55(Part 1): 1-8.

KOLPAKOVA, A.F. & KOLPAKOV, F.I. (1983) [Comparative study of sensitizing characteristics of the platinum group metals.] Gig. Tr. prof. Zabol., 7: 22-24 (in Russian).

KÖNIG, H.P. & HERTEL, R.F. (1990) [Search and identification of gaseous noble metal emissions from catalysts. Investigations into the hazardous potential of catalyst-born nobel metal emissions], Hanover, Fraunhofer Institute of Toxicology and Aerosol Research, 108 pp (Unpublished final research report to the Federal Ministry of Research and Technology) (in German).

KÖNIG, H.P., KOCK, H., & HERTEL, R.F. (1989) [Analytical determination of platinum with regard to the car catalyst issue.] In: 5th Colloquium on Trace Analysis by Atomic Absorption Spectrometry, Konstanz, 3-7 April, 1989], Überlingen, Perkin-Elmer GmbH, pp. 647-656 (in German).

KÖNIG, H.P., HERTEL, F.R., KOCH, W., & ROSNER, G. (in press) Determination of platinum emissions from a three-way catalyst equipped gasoline engine. Atmospheric environment.

KRAFT, L.A., POLIDORO, J.P., CULVER, R.M., & HAHN, D.W. (1978) Intravas device studies in rabbits: II. Effect on sperm output, fertility and histology of the reproductive tract. Contraception, 18: 239-251.

KRAL, H. & PETER, K. (1977) [Method for the production of platinum(II)-nitrate and its use for platinum catalysts for the catalytic after-burning of automobile exhaust gas], Münich, German Patent Office (Patent No. 2233677) (in German).

References

KRITSOTAKIS, K. & TOBSCHALL, H.J. (1985) [Trace determination of platinum by differential-pulse anodic stripping voltammetry (DPASV) using the glassy carbon electrode.] Fresenius Z. anal. Chem., **320**: 156-158 (in German).

LANGE, R.C., SPENCER, R.P., & HARDER, H.C. (1972) Synthesis and distribution of a radiolabeled antitumor agent: cis-diamminedichloroplatinum (II). J. nucl. Med., **13**: 328-330.

LAZNICKOVA, A., SEMECKY, V., LAZNICEK, M., ZUBER, V., KOKSAL, J., & KVETINA, J. (1989) Effect of oxoplatinum and CBDCA on renal functions in rats. Neoplasma, **36**: 161-169.

LECOINTE, P., MACQUET, J.P., BUTOUR, J.L., & PAOLETTI, C. (1977) Relative efficiencies of a series of square-planar platinum(II) compounds on Salmonella mutagenesis. Mutat. Res., **48**: 139-144.

LEE, D.S. (1983) Palladium and nickel in north-east Pacific waters. Nature(Lond.), **305**: 47-48.

LE HOUILLIER, R. & DE BLOIS, C. (1986) Alkyl cyanide medium for the determination of precious metals by atomic absorption spectrometry. Analyst, **111**: 291-294.

LELIEVELD, P., VAN DER VIJGH, J.F., VELDHUIZEN, R.W., VAN VELZEN, D., VAN PUTTEN, L.M., ATASSI, G., & DANGUY, A. (1984) Preclinical studies on toxicity, antitumour activity and pharmacokinetics of cisplatin and three recently developed derivatives. Eur. J. Cancer clin. Oncol., **20**: 1087-1104.

LEROY, A.F., WEHLING, M.L., SPONSELLER, H.L., FRIAUF, W.S., SOLOMON, R.E., & DEDRICK, R.L., (1977) Analysis of platinum biological materials by flameless atomic absorption spectrophotometry. Biochem. Med., **18**: 184-191.

LIFE SCIENCE RESEARCH (1980a) $Pt(NH_3)_4Cl_2$; assessment of its mutagenic potential in histidine auxotrophs of *Salmonella typhimurium*, Eye, Suffolk, Life Science Research, (Unpublished report No. 80/JOM001/005).

LIFE SCIENCE RESEARCH (1980b) $Pt(NH_3)_4Cl_2$; assessment of clastogenic action on bone marrow erythrocytes in the micronucleus test, Eye, Suffolk, Life Science Research (Unpublished report No. 80/JOM004/182).

LIFE SCIENCE RESEARCH (1980c) $Pt(NH_3)_4Cl_2$; assessment of its mutagenic potential in *Drosophila melanogaster*, using the sex-linked recessive lethal test, Eye, Suffolk, Life Science Research (Unpublished report No. 80/JOM002/260).

LIFE SCIENCE RESEARCH (1981a) Potassium chloroplatinate: assessment of clastogenic action on bone marrow erythrocytes in the micronucleous test, Eye, Suffolk, Life Science Research (Unpublished report No. 81/JOM013/241).

LIFE SCIENCE RESEARCH (1981b) TPC: assessment of its action on bone marrow cell chromosomes following sub-acute oral administration in the Chinese hamster, Eye, Suffolk, Life Science Research (Unpublished report No. 80/JOM008/497).

LIFE SCIENCE RESEARCH (1981c) Potassium chloroplatinate(II): assessment of its mutagenic potential in *Drosophila melanogaster*, using the sex-linked recessive lethal test, Eye, Suffolk, Life Science Research (Unpublished report No. 81/JOM012/474).

LIFE SCIENCE RESEARCH (1982) Potassium chloroplatinate: investigation of effects on bone marrow chromosomes of the Chinese hamster after sub-acute oral administration, Eye, Suffolk, Life Science Research (Unpublished report).

LINNETT, P.J. (1987) Platinum salt sensitivity. A review of the health aspects of platinum refining in South Africa. J. Mine Med. Off. Assoc. S. Afr., 63: 24-28.

LIPPERT & BECK (1983) [Platinum complexes in cancer therapy.] Chem. Zeit, 6: 190-199 (in German).

LIPPMAN, A.J., HELSON, C., HELSON, L., & KRAKOFF, I.H. (1973) Clinical trials of cis-diamminedichloroplatinum (NSC 119875). Cancer Chemother. Rep., 57: 191-200.

LITTERST, C.L. (1981) Alterations in the toxicity of cis-dichlorodiammineplatinum-II and in tissue localization of platinum as a function of NaCl concentration in the vehicle of administration. Toxicol. appl. Pharmacol., 61: 99-108.

LITTERST, C.L., TORRES, I.J., & GUARINO, A.M. (1976a) Plasma levels & organ distribution of platinum in the rat, dog & dogfish shark following single intravenous administration of cis-dichlorodiammineplatinum(II). J. clin. Hematol. Oncol., 7: 169-179.

LITTERST, C.L., GRAM, T.E., DEDRICK, R.L., LEROY, A.F., & GUARINO, A.M. (1976b) Distribution and disposition of platinum following intravenous administration of cis-diamminedichloroplatinum(II) (NSC 119875) to dogs. Cancer Res., 36: 2340-2344.

LITTERST, C.L., LEROY, A.F., & GUARINO, A.M. (1979) Disposition and distribution of platinum following parenteral administration of cis-dichlorodiammineplatinum-II to animals. Cancer Treat. Rep., 63: 1485-1492.

LO, F.B., ARAI, D.K., & NAZAR, M.A. (1987) Determination of platinum in urine by inductively coupled plasma atomic emission spectrometry. J. anal. Toxicol., 11: 242-246.

LOEBENSTEIN, J.R. (1982) Platinum-group metals. In: Minerals yearbook, centennial edition 1981. Vol. I. Metals and minerals, Washington, DC, US Department of the Interior, Bureau of Mines, pp. 667-678.

LOEBENSTEIN, J.R. (1988) Platinum-group metals. In: Minerals yearbook 1987. Vol. I. Metals and minerals, Washington, DC, US Department of the Interior, Bureau of Mines, pp. 689-700.

LOWN, B.A., MORGANTI, J.B., STINEMAN, C.H., D'AGOSTINO, R.B., & MASSARO, E.J. (1980) Tissue organ distribution and behavior effects of platinum following acute and repeated exposure of the mouse to platinum sulfate. Environ. Health Perspect., **34**: 203-212.

MCGAHAN, M.C. & TYCZKOWSKA, K. (1987) The determination of platinum in biological materials by electrothermal atomic absorption spectroscopy. Spectrochim. Acta, **42B**: 665-668.

MAESSEN, F.J.M., KREUNING, G., & BALKE, J. (1986) Experimental control of the solvent load of inductively coupled argon plasmas and effects of the chloroform plasma load on their analytical performance. Spectrochim. Acta, **41B**: 3-25.

MALANCHUK, M., BARKELEY, N., CONTNER, G., RICHARDS, M., SLATER, R., BURKART, J., & YANG, Y. (1974) Exhaust emissions from catalyst-equipped engines. In: Environmental Toxicology Research Laboratory, NERC, ed. Interim report: 1. Studies on toxicology of catalytic trace metal components. 2. Toxicology of automotive emissions with and without catalytic converters, Cincinnati, Ohio, Environmental Toxicology Research Laboratory, NERC, pp. 91-98.

MARCZENKO, Z. & KUS, S. (1987) Spectrophotometric determination of traces of platinum in palladium with dithizone after matrix precipitation as a compound with ammonia and iodide. Anal. Chim. Acta, **196**: 317-322.

MARONE, C., OLSINA, R., & SALAS, O. (1981) Spectrometric determination of platinum in a spent reforming catalyst. Cron. chim., **66**: 22-28.

MARSH, K.C., STERNSON, L.A., & REPTA, A.J. (1984) Post-column reaction detector for platinum(II) antineoplastic agents. Anal. Chem., **56**: 491-497.

MASON, B. (1966) Principles of geochemistry, 3rd ed., New York, John Wiley & Sons, 310 pp.

MASSARO, E.J., LOWN, B.A., MORGANTI, J.B., STINEMAN, C.H., & D'AGOSTINO, R.B. (1981) Sensitive biochemical and behavioral indicators of trace substance exposure: Part II. Platinum, Research Triangle Park, US Environmental Protection Agency, 3 pp (EPA-600/S1-81-015).

MASSMANN, W. & OPITZ, H. (1954) [On platinum allergy.] Zentralbl. Arbeitsmed. Arbeitsschutz, **4**: 1-4 (in German).

MATUSIEWICZ, H. & BARNES, R.M. (1983) Determination of metal chemotherapeutic agents (Al, Au, Li, and Pt) in human body fluids using electrothermal vaporization with inductively coupled plasma atomic emission spectroscopy. In: Chemical toxicology and clinical chemistry of metals, Oxford, United Kingdom, International Union of Pure and Applied Chemistry, pp. 49-52.

MERGET, R., SCHULTZE-WERNINGHAUS, G., MUTHORST, T., FRIEDRICH, W., & MEIER-SYDOW, J. (1988) Asthma due to the complex salts of platinum - a cross-sectional survey of workers in a platinum refinery. Clin. Allergy, **18**: 569-580.

MERGET, R., ZACHGO, W., SCHULTZE-WERNINGHAUS, G., BERGMANN, E.-M., BOLM-AUDORFF, U., FRIEDRICH, W., BURY, A.H., & MEIER-SYDOW, J. (1990) [Quantitative skin test and inhalative provocation test in asthma associated with platinum compounds.] Pneumologie, 44: 226-228 (in German).

MERRIN, C.A. (1976) A new method to prevent toxicity with high doses of cis-diammineplatinum (therapeutic efficacy in previously treated widespread and recurrent testicular tumors). In: Proceedings of the 12th Annual Meeting of the American Association for Cancer Research/American Society of Clinical Oncology, Toronto, 4-5 May, 1976, Chicago, Illinois, American Society of Clinical Oncology, p. 243 (ASCO Abstracts, No. C-26).

MICHAEL, L.W., FINELLI, V.N., ELIA, V.J., PETERING, H.G., LEE, S.D., & CAMPBELL, K.I. (1976) Interaction of metal binding agents from combustion products with trace metals. In: Proceedings of the Platinum Research Review Conference, Quail Roost Conference Center, Rougemont, North Carolina, 3-5 December 1975, Research Triangle Park, North Carolina, US Environmental Protection Agency, pp. 37A-53A.

MILLARD, H.T. (1987) Neutron activation determination of iridium, gold, platinum, and silver in geologic samples. J. radioanal. nucl. Chem., 113: 125-132.

MOJSKI, M. & KALINOWSKI, K. (1980) Extractive separation of platinum from macroamounts of palladium using triphenylphosphine oxide and its successive spectrophotometric determination by the stannous chloride method. Microchem. J., 25: 507-513.

MOLLMAN, J.E., HOGAN, W.M., GLOVER, D.J., & MCCLUSKEY, L.F. (1988) Unusual presentation of cis-platinum neuropathy. Neurology, 38: 488-490.

MOORE, W., Jr, HYSELL, D., CROCKER, W., & STARA, J. (1975a) Biological fate of a single adminstration of ^{191}Pt in rats following different routes of exposure. Environ. Res., 9: 152-158.

MOORE, W., Jr, HYSELL, D., HALL, L., CAMPBELL, K., & STARA, J. (1975b) Preliminary studies on the toxicity and metabolism of palladium and platinum. Environ. Health Perspect., 10: 63-71.

MOORE, W., Jr, MALANCHUK, M., CROCKER, W., HYSELL, D., COHEN, A., & STARA, J.F. (1975c) Whole body retention in rats of different ^{191}Pt compounds following inhalation exposure. Environ. Health Perspect., 12: 35-39.

MUELLER, B.J. & LOVETT, R.J. (1987) Salt-induced phase separation for the determination of metals as their diethyldithiocarbamate complexes by high-performance liquid chromatography. Anal. Chem., 59: 1405-1409.

MURDOCH, R.D. & PEPYS, J. (1984) Immunological responses to complex salts of platinum. I. Specific IgE antibody production in the rat. Clin. exp. Immunol., 57: 107-114.

References

MURDOCH, R.D. & PEPYS, J. (1985) Cross reactivity studies with platinum group metal salts in platinum-sensitised rats. Int. Arch. Allergy appl. Immunol., 77: 456-458.

MURDOCH, R.D. & PEPYS, J. (1987) Platinum group metal sensitivity: reactivity to platinum group metal salts in platinum halide salt-sensitive workers. Ann. Allergy, 59: 464-469.

MURDOCH, R.D., PEPYS, J., & HUGHES, E.G. (1986) IgE antibody responses to platinum group metals: a large-scale refinery survey. Br. J. ind. Med., 43: 37-43.

MYASOEDOVA, G.V., ANTOKOL'SKAYA, I.I., & SAVVIN, S.B. (1985) New chelating sorbents for noble metals. Talanta, 32: 1105-1112.

NAS (1977) Platinum-group metals, Washington, DC, National Academy of Science, 232 pp.

NEUMÜLLER, O.-A. (1987) [Römpps lexicon on chemistry], Stuttgart, W. Keller & Co., p. 3256 (in German).

NEWMAN-TAYLOR, A.J. (1981) Follow-up study of a group of platinum refinery workers, London, Brompton Hospital (Unpublished report).

NYGREN, O., VAUGHAN, G.T., FLORENCE, T.M., MORRISON, G.M., WARNER, I., & DALE, L.S. (1990) Determination of platinum in blood by adsorptive voltammetry. Anal. Chem., 62(15): 1637-1640.

PALLAS, J.E., Jr & JONES, J.B., Jr (1978) Platinum uptake by horticultural crops. Plant Soil, 50: 207-212.

PARISH, W.E. (1970) Short-term anaphylactic IgG antibodies in human sera. Lancet, 2: 591-592.

PARROT, J.L., HEBERT, R., SAINDELLE, A., & RUFF, F. (1969) Platinum and platinosis. Allergy and histamine release due to some platinum salts. Arch. environ. Health, 19: 685-691.

PEPYS, J. (1983) Allergy of the respiratory tract to low molecular weight chemical agents. In: Born, G.V.R., Farah, A., Herken, H., & Welch, A.D., ed. Handbook of experimental pharmacology, Berlin, Springer-Verlag, pp. 163- 185.

PEPYS, J., PICKERING, C.A.C., & HUGHES, E.G. (1972) Asthma due to inhaled chemical agents - complex salts of platinum. Clin. Allergy, 2: 391-396.

PEPYS, J., PARISH, W.E., CROMWELL, O., & HUGHES, E.G. (1979) Passive transfer in man and the monkey of type I allergy due to heat labile and heat stable antibody to complex salts of platinum. Clin. Allergy, 9: 99-108.

PERA, M.F., Jr & HARDER, H.C. (1977) Analysis of platinum in biological material by flameless atomic absorption spectrometry. Clin. Chem., 23: 1245-1249.

PICKERING, C.A.C. (1972) Inhalation tests with chemical allergens: complex salts of platinum. Proc. R. Soc. Med., 65: 272-274.

PINTO, A.L. & LIPPARD, S.J. (1985) Binding of the antitumour drug cis-diamminedichloroplatinum(II) (cisplatin) to DNA. Biochim. Biophys. Acta, 780: 167-180.

POTTER, N.M. & LANGE, W.H. (1981) Determination of noble metals and their distribution in automotive catalyst materials. Am. Lab., 1: 81-91.

PRIESNER, D., STERNSON, L.A., & REPTA, A.J. (1981) Analysis of total platinum in tissue samples by flameless atomic absorption spectrophotometry. Elimination of the need for sample digestion. Anal. Lett., 14: 1255-1268.

PURI, B.K., WASEY, A., KATYAL, M., & SATAKE, M. (1986) Spectrophotometric determination of platinum after extraction of its phenanthraquinone monoximate into molten naphtalene. Analyst, 111: 743-745.

RENNER, H. (1979)[a] [Platinum metals and alloys.] In: Bartholomé, E., Biekert, E., Hellmann, H., & Ley, H., ed. [Ullmann's encyclopedia of technical chemistry], 4th ed., Weinheim, Verlag Chemie, Vol. 18, pp. 697-705 (in German).

RENNER, H. (1984)[b] [Platinum metals.] In: Merian, E., Geldmacher-von-Mallinckrodt, M., Machata, G., Nürnberg, H.W., Schlipköter, H.W., & Stumm, W., ed. [Metals in the environment. Distribution, analysis, and biological relevance], Weinheim, Verlag Chemie, pp. 499-510 (in German).

RICKEY, F.A., SIMMS, P.C., & MUELLER, K.A. (1979) PIXE analysis of water with detection limits in the ppb range. IEEE Trans. nucl. Sci., NS-26(1): 1347-1351.

ROBERT, R.V.D., VAN WYK, E., & PALMER, R. (1971) Concentration of the noble metals by a fire-assay technique using sulphide as the collector, Johannesburg, South Africa, National Institute for Metallurgy (Report No. 1371).

ROBERTS, A.E. (1951) Platinosis. A five-year study of the effects of soluble platinum salts on employees in a platinum laboratory and refinery. Arch. ind. Hyg. occup. Med., 4: 549-559.

ROCKLIN, R.D. (1984) Determination of gold, palladium, and platinum at the parts-per-billion level by ion chromatography. Anal. Chem., 56: 1959-1962.

ROSENBERG, B. (1975) Possible mechanisms for the antitumor activity of platinum coordination complexes. Cancer Chemother. Rep., 59(Part 1): 589-598.

---

[a] English edition in preparation:
RENNER, H. (in press) Platinum metals, compounds and alloys. In: Ullmanns Encyclopaedia of Industrial Chemistry, 5th ed., Weinheim, VCH Verlagsgesellschaft.

[b] English edition in preparation:
RENNER, H. & SCHMUCKLER, G. (in press) Platinum-group metals. In: Merian, E., ed. Metals and their compounds in the environment, Weinheim, VCH Verlagsgesellschaft.

ROSENBERG, B. (1980) Clinical aspects of platinum anticancer drugs. In: Metal ions in biological systems, New York, Basel, Marcel Dekker, Inc., Vol. 12, pp. 127-196.

ROSENBERG, B. (1985) Fundamental studies with cisplatin. Cancer, 55: 2303-2316.

ROSENBERG, B., VAN CAMP, L., & KRIGAS, T. (1965) Inhibition of cell division in *Escherichia coli* by electrolysis products from a platinum electrode. Nature (Lond.), 205: 698-699.

ROSENBERG, B., RENSHAW, E., VAN CAMP, L., HARTWICK, J., & DROBNIK, J. (1967) Platinum-induced filamentous growth in *Escherichia coli*. J. Bacteriol., 93: 716- 721.

ROSHCHIN, A.V., VESELOV, V.G., & PANOVA, A.I. (1979) [Toxicology of platinum and platinum-group metals.] Gig. Tr. prof. Zabol., 9: 4-9 (in Russian).

ROSHCHIN, A.V., VESELOV, V.G., & PANOVA, A.I. (1984) Industrial toxicology of metals of the platinum group. J. Hyg. Epidemiol. Microbiol. Immunol., 28: 17-24.

ROSNER, G. & HERTEL, R.F. (1986) [Health risk assessment of platinum emissions from automotive exhaust gas catalysts.] Staub-Reinhalt. Luft, 46: 281-285 (in German).

ROSNER, G. & MERGET, R. (1990) Allergenic potential of platinum compounds. In: Dayan, A.D., Hertel, R.F., Heseltine, E., Kazantzis, G., Smith, E.M., & Van der Venne, M.T., ed. Immunotoxicity of metals and immunotoxicology, New York, London, Plenum Press, pp. 93-102.

ROSNER, G., KÖNIG, H.P., KOCH, W., KOCK, H., HERTEL, R.F., & WINDT, H. (1991) [Engine test stand experiments to assess platinum uptake by plants.] Angew. Bot., 65: 127-132 (in German).

RUFF, F., DI MATTEO, G., DUPUIS, J.P., HEBERT, R., & PARROT, J.-L. (1979) Réactions bronchopulmonaires et asthme dus au platine: incidences chez les ouvriers du platine en région parisienne. Rev. fr. Mal. Respir., 7: 206-208.

SAFIRSTEIN, R., DAYE, M., & GUTTENPLAN, J.B. (1983) Mutagenic activity and identification of excreted platinum in human and rat urine and rat plasma after administration of cisplatin. Cancer Lett., 18: 329-338.

SAINDELLE, A. & RUFF, F. (1969) Histamine release by sodium chloroplatinate. Br. J. Pharmacol., 35: 313-321.

SANDHU (1979) Evaluation of the mutagenic potential of platinum compounds, Durham, North Carolina Central University, 36 pp (EPA-600/1-79-033).

SCHACTER, L. & CARTER, S.K. (1986) The use of platinum to treat cancer. In: Precious metals: Proceedings of the 9th International Precious Metals Institute Conference, Lake Tahoe, Nevada, 9-12 June, 1986, Allentown, Pennsylvania, International Precious Metals Institute, pp. 359-367.

SCHLEMMER, G. & WELZ, B. (1986) Influence of the tube surface on atomization behavior in a graphite tube furnace. Fresenius Z. anal. Chem., 323: 703-709.

SCHLÖGL, R., INDLEKOFER, G., & OELHAFEN, P. (1987) [Micro particle emission from combustion engines with exhaust gas purification - X-ray photoelectron spectroscopy in environmental analysis.] Angew. Chem., 99: 312-322 (in German).

SCHULTZE-WERNINGHAUS, G., ROESCH, A., WILHELMS, O.-H., GONSIOR, E., & MEIER-SYDOW, J. (1978) [Bronchial asthma due to occupational allergy of immediate type (I) to platinum salts.] Dtsch. med. Wochenschr., 23: 972-975 (in German).

SCHULTZE-WERNINGHAUS, G., MERGET, R., ZACHGO, W., MUTHORST, T., MAHLESA, D., LISSON, R., & BOLM-AUDORFF, U. (1989) [Platinum salts as occupational allergens - a review.] Allergologie, 12: 152-157 (in German).

SCHUTYSER, P., GOVAERTS, A., DAMS, R., & HOSTE, J. (1977) Neutron activation analysis of platinum metals in airborne particulate matter. J. radioanal. Chem., 37: 651-660.

SHEARD, C. (1955) Contact dermatitis from platinum and related metals. Arch. Dermatol. Syphilol., 71: 357-360.

SHIMAZU, M. & ROSENBERG, B. (1973) A similar action to UV-irradiation and a preferential inhibition of DNA synthesis in *E. coli* by antitumor platinum compounds. J. Antibiot., 26: 243-245.

SIGHINOLFI, G.P., GORGONI, C., & MOHAMED, A.H. (1984) Comprehensive analysis of precious metals in some geological standards by flameless a.a. spectroscopy. Geostand. Newsl., 8: 25-29.

SIGSBY, J. (1976) Measurement of platinum from catalyst-equipped vehicles, combustion and attrition products. In: Proceedings of the Platinum Research Review Conference, Quail Roost Conference Center, Rougemont, North Carolina, 3-5 December 1975, Research Triangle Park, North Carolina, US Environmental Protection Agency, pp. 25A-31A.

SINGH, N. & GARG, A.K. (1987) Spectrophotometric determination of platinum(IV) with potassium butyl xanthate. Analyst, 112: 693-695.

SKINNER, P.E. (1989) Improvements in platinum plating. Platinum Met. Rev., 33(3): 102-105.

SKOGERBOE, R.K., HANAGAN, W.A., & TAYLOR, H.E. (1985) Concentration of trace elements in water samples by reductive precipitation. Anal. Chem., 57: 2815-2818.

SMITH, B.L., HANNA, M.L., & TAYLOR, R.T. (1984) Induced resistance to platinum in Chinese hamster ovary cells. J. environ. Sci. Health, A19: 267-298.

SPERNER, F. & HOHMANN, W. (1976) Rhodium-platinum gauzes for ammonia oxidation. Platinum Met. Rev., 20: 12-20.

STERNSON, L.A., REPTA, A.J., SHIH, H., HIMMELSTEIN, K.J., & PATTON, T.F. (1984) Disposition of cisplatin vs total platinum in animals and man. Dev. Oncol., 17: 126-137.

STOCKMAN, H.W. (1983) Neutron activation determination of noble metals in rock: a rapid radiochemical separation based on tellurium coprecipitation. J. radioanal. Chem., 78: 307-317.

STOKINGER, H.E. (1981) Platinum-group metals - platinum, Pt, palladium, Pd, iridium, Ir, osmium, Os, rhodium, Rh, ruthenium, Ru. In: Clayton, G.D. & Clayton, F.E., ed. Patty's industrial hygiene and toxicology, New York, John Wiley & Sons, pp. 1853-1871.

SURAIKINA, T.I., ZAKHAROVA, I.A., MASHKOVSKII, Y., & FONSHTEIN, L.M. (1979) Study of the mutagenic action of platinum and palladium compounds on bacteria. Cytol. Genet. (USSR), 13: 50-54 (Translation from: Tsitol. i. Genet. 13: 486-491).

SWIERENGA, S.H., GILMAN, J.P., & MCLEAN, J.R. (1987) Cancer risk from inorganics. Cancer Metastasis Rev., 6: 113-154.

TAUBLER, J.H. (1977) Allergic response to platinum and palladium complexes - determination of no-effect level, Research Triangle Park, North Carolina, US Environmental Protection Agency, Office of Research and Development, Health Effects Research Laboratories, pp. 1-81 (EPA/600/1-77-039).

TAYLOR, R.T. (1976) Comparative methylation chemistry of platinum, palladium, lead and manganese, Research Triangle Park, North Carolina, US Environmental Protection Agency, Office of Research and Development, Health Effects Research Laboratories, 25 pp (EPA/600/1-76-016).

TAYLOR, R.T. & HANNA, M.L. (1977) Methylation of platinum by methyl cobalamin. In: Drucker, H. & Wildung, R.E., ed. Biological implications of metals in the environment. Proceedings of the 15th Annual Hanford Life Science Symposium, Richland, Washington, 29 September - 1 October, 1975, Springfield, Virginia, US Department of Commerce, National Technical Information Service, pp. 36-51 (Energy Research and Development Administration Symposium Series, Vol. 42).

TAYLOR, R.T., HAPPE, J.A., HANNA, M., & WU, R. (1979) Platinum tetrachloride: mutagenicity and methylation with methylcobalamin. J. environ. Sci. Health, A14: 87-109.

THEOPOLD, H.-M., ZOLLNER, M., SCHORN, K., SPAHMANN, J., & SCHERER, H. (1981) [Tissue reactions with platinum-iridium electrodes.] Laryng.-Rhinol., 60: 534-537 (in German).

THOMPSON, B.C., RAYBURN, L.A., & HOLADAY, S.R. (1981) Detection of platinum in brain tissue by PIXE. IEEE Trans. nucl. Sci., NS-28: 1396-1397.

THOMPSON, J.J. & HOUK, R.S. (1986) Inductively coupled plasma mass spectrometric determination for multielement flow injection analysis and elemental speciation by reversed-phase liquid chromatography. Anal. Chem., 58: 2541-2548.

TILLERY, J.B. & JOHNSON, D.E. (1975) Determination of platinum, palladium, and lead in biological samples by atomic absorption spectrophotometry. Environ. Health Perspect., 12: 19-26.

TJIOE, P.S., VOLKERS, K.J., KROON, J.J., DE GOEIJ, J.J.M., & THE, S.K. (1984) Determination of gold and platinum traces in biological materials as a part of a multielement radiochemical activation analysis system. Int. J. environ. anal. Chem., 17: 13-24.

TOBE, M.L. & KHOKHAR, A.R. (1977) Structure, activity, reactivity and solubility relationships of platinum diammine complexes. J. clin. Hematol. Oncol., 7: 114-137.

TOBERT, A.J. & DAVIES, D.R. (1977) Effect of copper and platinum intrauterine devices on endometrial morphology and implantation in the rabbit. J. Reprod. Fertil., 50: 53-59.

TÖLG, G. & ALT, F. (1990) [Contribution to an improvement in extreme trace analysis of metallic platinum in biotic and environmental material.] In: [Abstract prepared for the Third Conference of the BMFT Study Group on Precious Metal Emissions, 15 October, 1990], Dortmund, Institute of Spectrochemistry, 7 pp (Unpublished report) (in German).

TSO, T.C., SOROKIN, T.P., & ENGELHAUPT, M.E. (1973) Effects of some rare elements on nicotine content of the tobacco plant. Plant Physiol., 51: 805-806.

VALENTE, I.M., MINSKI, M.J., & PETERSON, P.J. (1982) Neutron activation analysis of noble metals in vegetation. J. radioanal. Chem., 71: 115-127.

VAN DEN BERG, C.M.G. & JACINTO, G.S. (1988) The determination of platinum in sea water by adsorptive cathodic stripping voltammetry. Anal. Chim. Acta, 211: 129-139.

VAN DER VIJGH, W.J.F., ELFERINK, F., & POSTMA, G.J. (1984) Determination of ethylenediamineplatinum(II) malonate in infusion fluids, human plasma and urine by high-performance liquid chromatography. J. Chromatogr., 310: 335-342.

VENABLES, K.M., DALLY, M.B., NUNN, A.J., STEVENS, J.F., STEPHENS, R., FARRER, N., HUNTER, J.V., STEWART, M., HUGHES, E.G., & NEWMAN-TYLOR, A.J. (1989) Smoking and occupational allergy in workers in a platinum refinery. Br. Med. J., 299: 939-942.

VENITT, S., CROFTON-SLEIGH, C., HUNT, J., SPEECHLEY, V., & BRIGGS, K. (1984) Monitoring exposure of nursing and pharmacy personnel to cytotoxic drugs: urinary mutation assays and urinary platinum as markers of absorption. Lancet, January 14: 74-77.

VOGL, S.E., ZARAVINOS, T., & KAPLAN, B.H. (1980) Toxicity of cis-diamminedichloroplatinum II given in a two-hour outpatient regimen of diuresis and hydration. Cancer, 45: 11-15.

References

VON BOHLEN, A., ELLER, R., KLOCKENKÄMPER, R., & TÖLG, G. (1987) Microanalysis of solid samples by total-reflection X-ray fluorescence spectrometry. Anal. Chem., 59: 2551-2555.

VON HOFF, D.D., SCHILSKY, R., REICHERT, C.M., REDDICK, R.L., ROZENCWEIG, M., YOUNG, R.C., & MUGGIA, F.M. (1979) Toxic effects of cis-dichlorodiammine-platinum(II) in man. Cancer Treat. Rep., 63: 1527-1531.

VRANA, O., KLEINWÄCHTER, V., & BRABEC, V. (1983) Determination of platinum in urine by differential pulse polarography. Talanta, 30: 288-290.

WANG, T.C. & TAN, C.K. (1987) Photodegradation of trichloroethylene in microheterogeneous aqueous systems. Environ. Int., 13: 359-362.

WARD, J.M. & FAUVIE, K.A. (1976) The nephrotoxic effects of cis-diamminedichloroplatinum(II) (NSC-II 9875) in male F344 rats. Toxicol. appl. Pharmacol., 38: 535-547.

WARD, J.M., YOUNG, D.M., FAUVIE, K.A., WOLPERT, M.K., DAVIS, R., & GUARINO, A.M. (1976) Comparative nephrotoxity of platinum cancer chemotherapeutic agents. Cancer Treat. Rep., 60: 1675-1678.

WATERS, M.D., VAUGHAN, T.O., ABERNETHY, D.J., GARLAND, H.R., COX, C.C., & COFFIN, D.L. (1975) Toxicity of platinum (IV) salts for cells of pulmonary origin. Environ. Health Perspect., 12: 45-56.

WEAST, R.C. & ASTLE, M.J. (1981) CRC handbook of chemistry and physics, 62nd ed. Boca Raton, Florida, CRC Press, pp. B-128-B-129.

WELZ, B. & SCHLEMMER, G. (1987) Glassy carbon tubes for electrothermal atomic absorption spectrometry. Fresenius Z. anal. Chem., 327: 246-252.

WEMYSS, R.B., & SCOTT, R.H. (1978) Simultaneous determination of platinum group metals and gold in ores and related plant materials by inductively-coupled plasma-optical emission spectrometry. Anal. Chem., 50: 1694-1697.

WHITEHOUSE, M.W. & GARRETT, I.R. (1984) Heavy metal (Au, Pt) nephropathy: studies in normal and inflamed rats. In: Rainsford, K.D. & Velo, G.P., ed. Advances in inflammation research, New York, Raven Press, Vol. 6, pp 291-294.

WIELE, H. & KUCHENBECKER, H. (1974) [Spectrophotometric determination of rhenium in the presence of platinum in bimetallic catalysts.] Erdöl, Kohle, Erdgas, Petrochem., 27: 90-92 (in German).

WINDHOLZ, M. (1976) The Merck index, 9th ed., Rahway, New Jersey, Merck & Co., pp. 75, 977-991, 1114.

WOLFE, G.W. (1979) PIXE analysis of atmospheric gases. IEEE Trans. nucl. Sci., NS-26: 1389-1391.

WOOD, J.M., FANCHIANG, Y.-T., & RIDLEY, W.P. (1978) The biochemistry of toxic elements. Q. Rev. Biophys., 11: 467-479.

WÜRFELS, M., JACKWERTH, E., & STOEPPLER, M. (1987) [On the problem of disturbances of inverse voltammetric trace analysis after pressure decomposition of biological samples.] Fresenius Z. anal. Chem., 329: 459-461 (in German).

ZACHGO, W., MERGET, R., & SCHULTZE-WERNINGHAUS, G. (1985) [Proof of specific IgE against low molecular weight substances (platinum salts).] Atemweg. Lungenkr., 11: 267-268 (in German).

ZOLOTOV, YU.A., PETRUKHIN, O.M., MALOFEEVA, G.I., MARCHEVA, E.V., SHIRYAEVA, O.A., SHESTAKOV, V.A., MISKAR'YANTS, V.G., NEFEDOV, V.I., MURINOV, YU.I., & NIKITIN, YU.E. (1983) Determinations of platinum metals by X-ray fluorescence, atomic emission and atomic absorption spectrometry after preconcentration with a polymeric thioether. Anal. Chim. Acta, 148: 135-157.

RESUME

1. **Identité, propriétés physiques et chimiques et méthodes d'analyse**

   Le platine (Pt) est un métal noble, malléable et ductile de couleur blanc argent dont le numéro atomique est 78 et le poids atomique 195,9. On le trouve principalement à l'état naturel sous la forme des isotopes ^{194}Pt (32,9 %), ^{195}Pt (33,8 %) et ^{196}Pt (25,3 %). Son état d'oxydation maximum est de +6, les états +2 et +4 étant les plus stables.

   Le métal ne se corrode pas à l'air quelle que soit la température mais il peut être attaqué par les halogènes, les cyanures, le soufre et ses composés en fusion, les métaux lourds et les hydroxydes alcalins. L'attaque par l'eau régale ou par Cl_2/HCl (acide chlorhydrique concentré dans lequel on fait barboter du chlore) produit l'acide hexachloroplatinique, $H_2(PtCl_6)$, un important complexe du platine. Lorsqu'il est chauffé, le sel d'ammonium de l'acide hexachloroplatinique produit une substance grisâtre appelée "mousse de platine". La réduction d'une solution aqueuse d'acide hexachloroplatinique donne une poudre noire dispersée appelée "noir de platine".

   En solution aqueuse, les espèces chimiques dominantes sont des complexes. Beaucoup de ces sels complexes sont solubles dans l'eau en particulier ceux qui contiennent des ligands donneurs comme les halogènes ou l'azote. Le platine, comme les autres métaux du même groupe, ont une forte tendance à réagir sur les composés carbonés, en particulier les alcènes et les alcynes pour former des complexes de coordination de Pt(II).

   On dispose de diverses méthodes d'analyse pour le dosage du platine. La spectométrie d'absorption atomique et la spectroscopie d'émission en plasma sont très sélectives et spécifiques et constituent les méthodes de choix pour le dosage du platine dans les échantillons d'origine biologique ou environnementale. Dans différents milieux, on a obtenu avec ces méthodes des limites de détection de l'ordre du µg par kg ou par litre.

La spectroscopie d'émission atomique en plasma induite par haute fréquence (PIHF) est supérieure à l'absorption atomique électrothermique du fait que les effets de matrice sont plus faibles et qu'il y a possibilité d'exécuter simultanément l'analyse d'autres éléments.

2. Sources d'exposition humaine et environnementale

La concentration moyenne du platine dans la lithosphère, c'est-à-dire la croûte rocheuse de la terre, est estimée à 0,001-0,005 mg/kg. On trouve le platine soit à l'état natif (métallique) soit en combinaison, dans un certain nombre de minéraux. Les sources d'importance économique se trouvent en République d'Afrique du Sud et en URSS. La teneur en platine de ces gisements est de 1-500 mg/kg. Au Canada, les métaux du groupe du platine (platine, palladium, iridium, osmium, rhodium et ruthénium) sont présents dans les minerais sulfurés de cuivre et de nickel à une concentration moyenne de 0,3 mg/kg; l'affinage du cuivre et du nickel porte cette concentration à plus de 50 mg/kg. De petites quantités sont extraites de mines situées aux Etats-Unis d'Amérique, en Ethiopie, aux Philippines et en Colombie.

La production minière mondiale des métaux du groupe du platine, constituée à 40-50% de platine, augmente régulièrement depuis les deux dernières décennies. En 1971, la production était de 127 tonnes, dont 51-64 tonnes de platine. Depuis l'apparition du pot d'échappement catalytique, la production minière mondiale de ces métaux est passée à environ 270 tonnes (dont 108-153 tonnes de platine) en 1987. En 1989, la demande totale de platine dans le monde occidental était d'environ 97 tonnes.

La principale utilisation du platine tient à ses propriétés catalytiques exceptionnelles. Les autres applications industrielles sont basées sur d'autres propriétés remarquables de ce métal, en particulier sa résistance à la corrosion chimique dans un grand intervalle de température, son point de fusion élevé, sa ductilité et sa grande résistance mécanique. Le platine est également utilisé en joaillerie et en art dentaire.

Résumé

Certains complexes du platine, en particulier le cis-diamminedichloroplatine(II) ou cisplatine sont utilisés en thérapeutique.[a]

On ne dispose pas de données sur les émissions de platine d'origine industrielle. Lors de l'utilisation de catalyseurs à base de platine, une certaine quantité de métal peut s'échapper dans l'environnement selon le type de catalyseur. Parmi les catalyseurs fixes utilisés dans l'industrie, seuls ceux qui servent à l'oxydation de l'ammoniac dégagent des quantités importantes de platine.

Les catalyseurs de véhicules automobiles constituent des sources mobiles de platine. Selon des données limitées, l'attrition des anciens catalyseurs en granulés se situe entre 0,8 et 1,9 µg/par km parcouru. Environ 10% de ce platine est soluble dans l'eau.

Les résultats fournis par des mesures au banc d'essai montrent que les pots catalytiques à trois voies utilisant des catalyseurs monolithiques de la nouvelle génération réduisent les émissions totales de platine d'un facteur de 100-1000 par rapport aux catalyseurs en granulés. A des vitesses simulées de 60, 100 et 140 km/heure on a constaté que l'émission totale de platine se situait entre 3 et 39 ng/m^3 dans les gaz d'échappement, ce qui correspond environ à 2-39 ng par kilomètre parcouru. Le diamètre aérodynamique moyen des particules émises variait, lors des différents essais, entre 4 et 9 µm. D'après quelques indices, on peut penser que la majeure partie du platine est émise sous forme de métal ou de particules oxydées en surface.

---

[a] Cette monographie traite spécialement du platine et de certains de ses dérivés importants du point de vue professionnel ou écologique. Une étude détaillée des effets toxiques du cisplatine en tant que médicament anti-cancéreux et de ses analogues chez l'homme et l'animal sortirait du cadre de cette série car il s'agit de produits utilisés essentiellement comme agents thérapeutiques. En outre leurs propriétés toxiques sont exceptionnelles comparées à celles des autres dérivés du platine.

3. Transport, répartition et transformation dans l'environnement

Les métaux du groupe du platine sont rares dans le milieu ambiant, comparativement aux autres éléments. Dans les zones très industrialisées, on peut trouver d'importantes quantités de platine dans les sédiments des cours d'eau. On pense que les matières organiques, par exemple les acides humiques et fulviques, se lient au platine, cette réaction étant sans doute facilitée par des conditions convenables de pH et de potentiel redox dans le milieu aquatique.

Dans le sol, la mobilité du platine dépend du pH, du potentiel redox, de la teneur en chlore de l'eau qui imprègne le sol et de l'état naturel du platine dans les roches primaires. On estime que le platine n'est mobilisé que dans des conditions d'acidité extrême ou lorsque l'eau du sol est très riche en chlore.

On a montré qu'*in vitro* certains complexes du platine(IV) pouvaient être méthylés en présence de platine(II) par la méthylcobalamine bactérienne dans des conditions abiotiques.

4. Concentrations dans l'environnement et exposition humaine

La base de données relatives aux concentrations dans l'environnement est très limitée en raison de la très faible teneur de celui-ci en platine et des problèmes d'analyse que cela pose.

Les concentrations de platine dans des échantillons d'air ambiant prélevés à proximité d'autoroutes aux Etats-Unis d'Amérique avant l'introduction du pot catalytique se situaient en dessous de la limite de détection de 0,05 pg/m^3. Un certain nombre de données récentes en provenance d'Allemagne indiquent qu'à proximité des routes, des concentrations de platine dans l'air ambiant (échantillons de matière particulaire) vont de moins de 1 pg/m^3 à 13 pg/m^3. Dans les zones rurales, ces concentrations étaient du même ordre de grandeur (moins de 0,6 à 1,8 pg/m^3).

Résumé

A proximité immédiate des routes, les concentrations de platine dans l'air ambiant qui résultent de l'introduction de catalyseurs en granulés ont été évaluées à partir de modèles de dispersion et sur la base des données expérimentales relatives aux émissions. Etant donné que l'émission totale de platine d'un catalyseur de type monolithique est plus faible, sans doute d'un facteur allant de 100 à 1000, que celle d'un catalyseur en granulés, les concentrations en platine provenant de ce type de catalyseur devraient être de l'ordre du picogramme au femtogramme par mètre cube.

En divers endroits de Californie, on a trouvé, dans la poussière déposée sur les plantes à larges feuilles, des concentrations de 37-680 μg/kg de poids sec. Le nombre d'échantillons était limité mais les résultats montrent tout de même que le pot catalytique libère du platine dans l'environnement immédiat des routes.

Des cultures de graminées ont été exposées dans des serres expérimentales pendant quatre semaines à des gaz d'échappement légèrement dilués provenant d'un moteur équipé d'un catalyseur à trois voies (vitesse simulée 100 km/heure): à la limite de détection de 2ng par gramme de poids sec, on n'a pas trouvé de platine.

Des analyses effectuées sur les sédiments du lac Michigan ont montré que du platine s'y était déposé depuis une cinquantaine d'années à un rythme assez uniforme. Des concentrations dans des carottes de 1 à 20 cm étaient comprises entre 0,3 et 0,4 μg/kg de poids sec seulement.

On ne signale pas la présence de platine dans les eaux douces, en revanche de fortes concentrations (730 à 31 220 μg/kg de poids sec) ont été mesurées dans les sédiments d'un canal très pollué du Rhin en Allemagne.

Dans des échantillons de bois de *Pinus flexilis* on a trouvé des concentrations de platine allant de 0 (non décelable) à 56 μg/kg (poids des cendres). Toutefois la teneur du sol voisin était du même ordre et ces données plutôt limitées n'indiquent aucune tendance à l'accumulation.

Dans des échantillons isolés de végétaux provenant d'un sol extrêmement basique, on a mesuré des teneurs en platine allant de 100 à 830 μg/kg (poids sec).

Dans des échantillons d'eau de mer, on a relevé des concentrations allant de 37 à 332 pg/litre. Des carottes de sédiment prélevées dans le Pacifique oriental présentaient des teneurs en platine allant de 1,1 à 3 µg/kg (poids sec). La concentration la plus élevée (21,9 µg/kg) a été mesurée dans des sédiments océaniques à distance du littoral. Les algues macroscopiques marines présentent des teneurs en platine allant de 0,08 à 0,32 µg/kg de poids sec.

Dans la population générale le taux de platine sanguin se situe entre 0,1 et 2,8 µg/litre. Dans le sérum de travailleurs exposés au platine de par leur activité professionnelle, on a relevé des concentrations de 150 à 440 µg/litre.

La base de données relative aux concentrations de platine sur les lieux de travail est limitée. En raison de problèmes d'analyse, les données anciennes (0,9 à 1700 µg/m^3) ne sont probablement pas fiables. Toutefois on peut déduire de ces données que l'exposition aux sels de platine était à l'époque plus forte que la limite d'exposition professionnelle de 2 µg/m^3 qui est actuellement en vigueur dans la plupart des pays. Des études récentes effectuées sur les lieux de travail font état de concentrations qui sont, soit inférieures à la limite de détection de 0,05 µg/m^3, soit comprises entre 0,08 et 0,1 µg/m^3.

5. Cinétique et métabolisme

Une seule exposition de 48 minutes par la voie respiratoire à du platine sous différentes formes chimiques (5-8 mg/m^3) a montré que le ^{191}Pt inhalé était rapidement éliminé de l'organisme. On observe ensuite une phase d'élimination plus lente au cours de la période suivant l'exposition. Dix jours après exposition à du ^{191}PtCl$_4$, du ^{191}Pt(SO$_4$), du ^{191}PtO$_2$ et du ^{191}Pt sous forme métallique, la rétention du ^{191}Pt dans l'ensemble de l'organisme était respectivement de 1, 5, 8 et 6% de la dose initiale. La majeure partie du ^{191}Pt a été éliminée des poumons par l'action de l'ascenseur mucociliaire puis avalée et excrétée dans les matières fécales (temps de demi-élimination, 24 h.). Une petite fraction du ^{191}Pt a été décelée dans les urines, ce qui indique que la

Résumé

résorption est très faible au niveau des poumons et des voies digestives.

Lors d'une étude sur la destinée comparée du $^{191}PtCl_4$ administré à des rats selon différentes voies à raison de 25uCi par animal, on a constaté que c'était la voie intraveineuse qui entraînait la rétention la plus forte, suivie par la voie intratrachéenne. Elle était minimale après administration par voie orale. Etant donné que seule une très faible partie du produit administré par voie orale a été résorbée, l'essentiel a traversé les voies digestives et a été excrété dans les matières fécales. Au bout de trois jours, on ne décelait plus dans l'ensemble de l'organisme que moins de 1% de la dose initiale. Après administration intraveineuse, le ^{191}Pt se retrouvait en quantités pratiquement égales dans les matières fécales et dans l'urine. L'élimination était plus faible qu'après administration orale. Au bout de cette même période le taux de rétention dans l'ensemble de l'organisme était d'environ 65% et au bout de 28 jours il se situait encore à 14% de la dose initiale. A titre de comparaison, au bout de ces deux intervalles de temps environ 22 et 8% respectivement de la dose initiale demeuraient dans l'organisme après administration intratrachéenne.

Les principaux sites d'accumulation sont les reins, le foie, la rate et les glandes surrénales. La forte quantité de ^{191}Pt retrouvée dans les reins montre qu'une fois absorbé, le platine s'accumule en majeure partie dans ces organes d'où il est excrété dans l'urine. La quantité plus faible trouvée dans le cerveau montre que les ions platine ne traversent qu'en faible proportion la barrière hémo-méningée.

Contrairement aux sels solubles dans l'eau, le PtO_2, qui est insoluble, n'a été résorbé qu'en quantités très faibles, même après administration dans la nourriture à dose très élevée, c'est-à-dire correspondant à une dose totale de platine de 4308 mg par rat sur une période de quatre semaines.

Qu'il s'agisse des sels simples ou du cisplatine, il est établi que l'élimination se fait en deux phases : une phase initiale rapide suivie d'une phase prolongée au cours de la période suivant l'exposition, et que rien ne permet de penser que les modalités de rétention soient très différentes. Toutefois le cisplatine est très stable dans les liquides extracellulaire en raison de la forte concentration en ions chlorure qui suppriment l'hydratation. Ainsi s'explique que cette substance soit excrétée presque entièrement sous forme inchangée. Contrairement au cas des sels simples, elle est excrétée principalement dans les urines.

6. Effets sur les mammifères de laboratoire et les systèmes d'épreuve *in vitro*

La toxicité aiguë du platine est principalement fonction de la forme sous laquelle il se trouve. Les sels solubles sont beaucoup plus toxiques que les sels insolubles. Par exemple la toxicité par voie orale pour les rats (DL_{50}) décroît dans l'ordre suivant: $Na_2[PtCl_6]$ (25-50 mg/kg) > $(NH_4)_2[PtCl_6]$ (195-200 mg/kg) > $PtCl_4$ (240 mg/kg) > $Pt(SO_4)_2.4H_2O$ (1010 mg/kg) > $PtCl_2$ (> 2000 mg/kg) > PtO_2 (> 8000 mg/kg). En ce qui concerne ces deux derniers composés on n'a pas pu calculer la valeur de la DL_{50}.

Des tests cutanés pratiqués sur des lapins albinos ont montré que PtO_2, $PtCl_2$, $K_2[PtCl_4]$, $[Pt(NO_2)_2(NH_3)_2]$, $Pt(C_5H_7O_2)_2$ et *trans*-$[PtCl_2(NH_3)_2]$ pouvaient être considérés comme non irritants. Par contre $(NH_4)_2[PtCl_6]$, $(NH_4)_2[PtCl_4]$, $Na_2[PtCl_6]$, $Na_2[Pt(OH)_6]$, $K_2[Pt(CN)_4]$, $[Pt(NH_3)_4]Cl_2$, et *cis*-$[PtCl_2(NH_3)_2]$ se sont révélés irritants mais à des degrés divers.

Des tests d'irritation oculaire ont montré que tous les composés testés avaient une action irritante. Le *trans*-$[PtCl_2(NH_3)_2]$ ainsi que $(NH_4)_2[PtCl_4]$ se sont révélés corrosifs.

De très sérieuses difficultés respiratoires ont été observées après injection intraveineuse de complexes chloroplatiniques à des cobayes et à des rats, vraisemblablement par suite d'une libération d'histamine d'origine non allergique. Cette libération aspécifique d'histamine complique l'interprétation des études sur l'animal et sur l'homme en ce qui concerne le diagnostic de la sensibilisation allergique. Après injection sous-

Résumé

cutanée et intraveineuse de $Pt(SO_4)_2$, trois fois par semaine pendant quatre semaines, on n'a pas constaté l'apparition d'un état allergique, à en juger d'après les résultats des épreuves cutanées effectuées sur des cobayes et des lapins, le transfert passif et les tests sur le coussinet plantaire de la souris. L'administration d'un complexe platine/albumine d'oeuf n'a pas non plus entraîné de sensibilisation chez les animaux de laboratoire.

On a essayé sans succès de sensibiliser des rats Lister femelles avec du tétrachloroplatinate d'ammonium, $(NH_4)_2[PtCl_4]$ par administration intrapéritonéale, intramusculaire, intradermique, sous-cutanée, intratrachéenne, et dans le coussinet plantaire en présence de *Bordetella pertussis* comme adjuvant; la sensibilisation a été évaluée par un test cutané direct, un test d'anaphylaxie cutanée passive (PCA) ou un test avec radio-allergo-absorbant (RAST). Toutefois le test PCA a donné un résultat positif après administration de conjugués platine/protéines.

Chez des singes Cynomolgus (*Macaca fascicularis*) exposés à de l'hexachloroplatinate de sodium, $Na_2[PtCl_6]$ exclusivement en inhalations nasales à la dose de 200 µg par m³, 4 h par jour, deux fois par semaine pendant 12 semaines, on a observé un déficit pulmonaire sensiblement plus élevé que chez les témoins. Dans le cas de l'hexachloroplatinate d'ammonium, il a fallu exposer les animaux simultanément à de l'ozone (200 µ/m³) pour obtenir une hypersensibilisation cutanée et une hyperréactivité pulmonaire significatives.

Des études sur des rats Sprague-Dawley mâles ont montré que des sels comme $PtCl_4$ (182 mg/litre d'eau de boisson) et comme $Pt(SO_4).4H_2O$ (248 mg/litre) n'affectaient pas le gain de poids au cours de la période d'observation de 4 semaines. En triplant la concentration de platine, il y a eu une réduction de 20% du gain de poids, mais seulement pendant la première semaine, parallèlement à une diminution de 20% de la prise de nourriture et de boisson.

On ne dispose que de données expérimentales limitées à propos des effets du platine sur la reproduction, et plus particulièrement à propos de ses éventuels effets embryotoxiques et tératogènes. Le $Pt(SO_4)_2$ (200 mg Pt/kg) a provoqué la mise bas de souriceaux Swiss ICR de poids réduit du jour 8 au jour 45 du post-partum. Le principal effet de $Na_2[PtCl_6]$ (20 mg Pt/kg) a consisté dans une réduction de l'activité de la progéniture lorsque les mères avaient été exposées le douzième jour de la gestation. Les fils et les feuilles de platine sont considérés comme inertes biologiquement et les effets nocifs constatés après implantation dans l'utérus de rattes et de lapines étaient probablement dus à la présence physique d'un corps étranger.

Après administration à des rattes gravides d'une dose de ^{191}Pt égale à 25 microcuries par animal, le 18ème jour de la gestation, on a constaté un passage limité à travers la barrière foeto-placentaire.

Plusieurs dérivés du platine se sont révélés mutagènes dans un certain nombre de systèmes bactériens. Lors d'études comparatives on a constaté que la mutagénicité du cisplatine était plusieurs fois supérieure à celle des autres composés. Des études *in vitro* ont montré que, dans le système cellulaire mammalien CHO-HGPT, l'activité mutagène relative s'établissait selon la proportion 100:9:0,3 respectivement pour les composés suivants: *cis*-$[PtCl_2(NH_3)_2]$, $K[PtCl_3(NH_3)]$, et $[Pt(NH_3)_3Cl]Cl$. La mutagénicité de $K_2[PtCl_4]$ et du *trans*-$[PtCl_2(NH_3)]$ était marginale, tandis que le $[Pt(NH_3)_4]Cl_2$ n'était pas mutagène. Les composés $K_2[PtCl_4]$ et $[Pt(NH_3)_4]Cl_2$ ne l'étaient pas non plus dans les tests suivants: mutation récessive léthale liée au sexe chez *Drosophila melanogaster*, recherche de micronoyaux dans des cellules de souris et le test sur moelle osseuse de hamster. On ne possède de données expérimentales sur la cancérogénicité des dérivés du platine que dans le cas du cisplatine pour lequel les preuves d'une activité cancérogène chez l'animal sont suffisantes. Cependant le cisplatine et ses analogues font plutôt figure d'exception si on les compare aux autres dérivés du platine. Cela transparaît dans leur activité antitumorale, dont le mécanisme est très particulier. On pense que celle-ci est, semble-t-il, due à la formation de ponts intercaténaires qui ne se produit qu'en présence de l'isomère *cis* et pour une certaine position de la guanine. Les cellules tumorales ne sont plus en mesure de se répliquer, alors que les cellules normales conservent leur capacité de réplication après avoir réparé les lésions provoquées par le cisplatine.

Résumé

7. Effets sur l'homme

L'exposition aux sels de platine se limite essentiellement aux ambiances de travail, et plus précisément aux ateliers d'affinage du platine et aux unités de production de catalyseurs.

Les composés principalement responsables de l'hypersensibilité aux sels de platine[a] sont l'acide hexachloroplatinique $H_2[PtCl_6]$ et un certain nombre de sels comme l'hexachloroplatinate d'ammonium $(NH_4)_2[PtCl_6]$, le tétrachloroplatinate de potassium, $K_2[PtCl_4]$, et le tétrachloroplatinate de sodium, $Na_2[PtCl_4]$. Les complexes dans lesquels il n'y a pas d'halogènes coordonnés au platine (complexes non halogénés), comme $K_2[Pt(NO_2)_4]$, $[Pt(NH_3)_4]Cl_2$ et $[Pt\{(NH_2)_2CS\}_4]Cl_2$, de même que les complexes neutres comme le cis-$[PtCl_2(NH_3)]$, ne sont pas allergéniques car ils réagissent avec les protéines pour former un antigène complet.

Les symptômes de cette hypersensibilité sont les suivants: urticaire, dermatite de contact, ainsi qu'un certain nombre de troubles respiratoires, comme reniflement, essoufflement, cyanose et asthme grave. La période de latence entre le premier contact avec des sels de platine et l'apparition des symptômes dure de quelques semaines à plusieures années. Après sensibilisation les symptômes ont tendance à s'aggraver aussi longtemps que les travailleurs sont exposés sur leur lieu de travail, mais ils disparaissent, en général, dès que cesse l'exposition. Toutefois, si une exposition de longue durée fait suite à la sensibilisation, les symptômes risquent de ne jamais disparaître complètement.

Bien qu'il soit impossible de tirer des données publiées une relation dose-effet qui ne soit pas ambiguë, il semble que le risque d'apparition d'une hypersensibilité aux sels de platine soit en corrélation avec l'intensité de l'exposition. Le platine métallique ne

---

[a] Le terme platinose n'est plus usité pour désigner les affections provoquées par les sels de platine, car il implique une fibrose pulmonaire chronique du type silicose. Il est préférable d'utiliser le terme allergie aux sels de platine, ou allergie aux composés du platine contenant des ligands halogénés réactifs ou mieux, hypersensibilité aux sels de platine.

semble pas être allergénique. A l'exception d'un cas unique de dermatite de contact, aucune réaction allergique n'a été signalée.

Les manifestations cliniques de l'hypersensibilité aux sels de platine sont celles d'une véritable réaction allergique. Le mécanisme de cette réaction est du type 1 (médiation par les IgE). Sur la base d'épreuves *in vivo* et *in vitro* on pense que chez les sujets sensibles il se forme des anticorps IgE dirigés contre les complexes chloroplatiniques. Les sels de platine de faible masse moléculaire relative se comportent comme des haptènes qui se combinent aux protéines pour former des antigènes complets.

Les tests cutanés avec des sels de platine dilués permettent de surveiller les réactions allergiques de manière reproductible, fiable, assez sensible et très spécifique. Pour les contrôles de routine en milieu professionnel on utilise les composés suivants: $(NH_4)_2[PtCl_6]$, $Na_2[PtCl_6]$ et $Na_2[PtCl_4]$. Il n'existe pas d'épreuve *in vitro* dont la sensibilité et la fiabilité approchent celles des tests cutanés. Des tests immuno-enzymatiques ou par immunoallergosorption ont permis de mettre en évidence des anticorps IgE spécifiques dirigés contre les complexes chlorés du platine. Il y avait corrélation avec les résultats des tests cutanés mais l'utilisation du test RAST reste problématique en raison de son manque de spécificité.

Les tests cutanés et l'épreuve RAST montrent qu'il n'existe qu'une faible réactivité croisée entre les sels de platine et les sels de palladium. Des réactions d'hypersensibilité à d'autres métaux du groupe du platine ont également été observées, mais seulement chez des personnes allergiques aux sels de platine.

Le tabagisme, l'atopie et l'hyperréactivité pulmonaire aspécifique ont été associés à l'hypersensibilité aux sels de platine et pourraient être des facteurs prédisposants.

En ce qui concerne la population générale, on manque de données sur l'exposition effective dans les pays où le pot catalytique est devenu obligatoire. Les concentrations dans l'air ambiant estimées d'après de nouvelles données

Résumé

sur les émissions et sur la base de modèles de dispersion sont probablement inférieures d'au moins un facteur 10 000 à la limite d'exposition professionnelle de 1 mg/m³ adoptée par certains pays pour le platine total inhalable sous forme de poussières. Etant donné que le platine est très probablement présent dans les émissions sous forme métallique, le potentiel de sensibilisation du platine émis par les pots catalytiques est probablement très faible. Même si une partie du platine émis est soluble et potentiellement allergénique, la marge de sécurité par rapport à la limite d'exposition professionnelle pour les sels solubles de platine (2 µg/m³) serait d'au moins 2000.

Lors d'une étude immunologique préliminaire, on a pratiqué des tests cutanés sur trois volontaires au moyen d'extraits de matières particulaires émises par des véhicules à moteur. On n'a pas observé de réponse positive.

On ne dispose d'aucune donnée sur les risques de cancérogénicité pour l'homme attribuables au platine et à ses sels. Pour ce qui est du cisplatine les preuves de cancérogénicité sont jugées insuffisantes.

8. Effets sur d'autres organismes au laboratoire et dans la nature

Les complexes simples du platine ont des effets bactéricides. En observant que les complexes neutres comme le cisplatine inhibaient sélectivement la division cellulaire sans réduction de la croissance chez diverses bactéries gram-positives mais aussi, et surtout, gram-négatives on a eu l'idée de les utiliser comme agents anticancéreux.

On a observé qu'au sein d'un "microcosme" de laboratoire, la croissance des algues vertes du genre euglène était inhibée en présence d'acide hexachloroplatinique soluble aux concentrations de 250, 500, et 750 µg/litre. Le cisplatine a provoqué une chlorose et un ralentissement de la croissance chez la jacinthe d'eau *Eichornia crassipes* à la concentration de 2,5 mg/litre.

Après 3 semaines d'exposition à de l'acide hexachloroplatinique, $H_2[PtCl_6]$, on a observé chez la daphnie une mortalité correspondant à une CL_{50} de 520 µg de Pt par litre. Aux concentrations de 14 et 82 µg/litre, la reproduction (nombre de jeunes daphnies) était réduite dans la proportion de 16 et 50% respectivement.

Après exposition de saumons (*Oncorhyncus kisutch*) à de l'acide tétrachloroplatinique pendant une brève période de temps dans des conditions statiques, on a observé que les valeurs de la CL_{50} étaient respectivement égales à 15,5, 5,2 et 2,5 mg Pt/litre au bout de 24-, 48-, et 96 h. On constatait, à la dose de 0,3 mg/litre, une diminution globale de l'activité natatoire et du mouvement des opercules. Aux concentrations supérieures à cette valeur, des lésions apparaissaient au niveau des branchies et de l'organe olfactif. Les concentrations de 0,03 et 0,1 mg/litre étaient sans effet.

Toutes les études consacrées aux effets du platine sur les plantes terrestres concernent uniquement les chlorures solubles. A des concentrations allant de 3.10^{-5} à 15.10^{-5} mol/kg (5,9-29,3 mg/kg) il y a eu inhibition de la croissance de plants de haricots et de tomates en sol sableux. Neuf variétés horticoles en culture hydroponique ont présenté une réduction de leur poids à sec après adjonction de tétrachlorure de platine aux concentrations respectives de 0,057, 0,57, et 5,7 mg Pt/litre; il s'agissait de tomates, de poivrons, de fanes de navets et pour la concentration la plus élevée, de radis. A cette concentration les bourgeons et les feuilles immatures devenaient chlorotiques chez la plupart des espèces. En revanche, chez certaines espèces, le tétrachlorure de platine stimulait la croissance. Par ailleurs, la teneur la plus élevée en platine supprimait la transpiration, sans doute par accroissement de la résistance des stomates. On constatait également une stimulation de la croissance aux faibles teneurs en platine (0,5 mg/litre), lorsque l'on ajoutait au milieu nutritif d'une graminée sud-africaine (*Setaria vertillata*) du tétrachloroplatinate de potassium. Au bout de deux semaines, la racine la plus longue avait poussé de 65%. A la concentration utilisée, soit 2,5 mg Pt/litre, on observait des effets phytotoxiques tels que rabougrissement des racines et chlorose foliaire.

RESUMEN

1. Identidad, propiedades físicas y químicas, métodos analíticos

El platino (Pt) es un metal noble maleable, dúctil, de color plateado blanquecino; su número atómico es 78 y su peso atómico 195,09. Sus isótopos naturales más abundantes son ^{194}Pt (32,9%), ^{195}Pt (33,8%), y ^{196}Pt (25,3%). En los compuestos de platino, el estado de oxidación máxima es +6; los estados +2 y +4 son los más estables.

Aunque el metal no se corroe en el aire a ninguna temperatura, es sensible a los halógenos, los cianuros, el azufre, los compuestos de azufre fundentes, los metales pesados y los hidróxidos de álcalis. La digestión con agua regia o Cl_2/HCl (ácido clorhídrico concentrado por el que se burbujea cloro) produce ácido hexacloroplatínico, $H_2[PtCl_6]$, un importante complejo de platino. Cuando se calienta, la sal amónica del ácido hexacloroplatínico produce una esponja gris de platino. La reducción en solución acuosa produce un polvo dispersivo de color negro ("negro de platino").

Las propiedades químicas de los compuestos del platino en solución acuosa se ven dominadas por los compuestos complejos. Muchas de las sales, particularmente las que llevan ligandos donadores de halógeno o de nitrógeno, son solubles en agua. El platino, al igual que los otros metales de su grupo, tiene una pronunciada tendencia a reaccionar con los compuestos del carbono, especialmente los alquenos y los alquinos, formando complejos de coordinación Pt(II).

Existen diversos métodos analíticos para la determinación del platino. La espectrometría de absorción atómica (EAA) y la espectroscopia de emisión de plasma son sumamente selectivas y específicas y constituyen el método de elección para analizar el platino presente en muestras biológicas y medioambientales. Con esos métodos se han alcanzado en diversos medios límites de detección del orden de unos cuantos µg/kg o µg/litro.

La espectroscopia de emisión atómica con plasma de argón acoplado por inducción es preferible a la EAA electrotérmica por sus menores efectos matriciales y por la posibilidad de analizar simultáneamente muchos elementos.

2. Fuentes de la exposición humana y ambiental

Se calcula que la concentración media de platino en la litosfera o corteza terrestre es del orden de 0,001-0,005 mg/kg. El platino se encuentra en forma metálica o en varias formas minerales. Existen fuentes económicamente importantes en la República de Sudáfrica y en la URSS. El contenido de platino de esos depósitos es de 1-500 mg/kg. En el Canadá, los metales del grupo del platino (platino, paladio, iridio, osmio, rodio, rutenio) se encuentran en menas de sulfuro de cuproníquel con una concentración media de 0,3 mg/kg, pero esa concentración supera los 50 mg/kg durante el afinado del cobre y el níquel. En los EE.UU., Etiopía, Filipinas y en Columbia se extraen pequeñas cantidades.

La producción minera mundial de metales del grupo del platino, de la cual el 40-50% corresponde al platino, ha aumentado uniformemente durante los últimos 20 años. En 1971, la producción fue de 127 toneladas (51-64 toneladas de platino). A raíz de la introducción del catalizador de los gases de escape en los automóviles, la producción minera mundial de metales del grupo del platino aumentó hasta aproximadamente 270 toneladas (108-135 toneladas de platino) en 1987. En 1989, la demanda total de platino en el mundo occidental fue de unas 97 toneladas.

El uso principal del platino deriva de sus excepcionales propiedades catalíticas. Las demás aplicaciones industriales aprovechan otras notables propiedades, en particular la resistencia a la corrosión química en un amplio intervalo de temperaturas, su elevado punto de fusión, su gran resistencia mecánica y su buena ductilidad. El platino se usa asimismo en joyería y odontología.

Resumen

Ciertos complejos de platino, en particular el *cis*-diaminodicloroplatino(II) (cisplatino), tienen aplicaciones terapéuticas.[a]

No se dispone de datos sobre las emisiones de platino al medio ambiente a partir de fuentes industriales. El uso de catalizadores con platino puede entrañar la liberación de ese elemento al medio ambiente, según el tipo de catalizador. De los catalizadores estacionarios utilizados en la industria, sólo los empleados para la oxidación del amoniaco emiten cantidades significativas de platino.

Los catalizadores utilizados en automoción son fuentes móviles de platino. Se dispone de datos limitados que indican que el desgaste de platino a partir del antiguo catalizador en pastilla es de 0,8 a 1,9 μg por km recorrido. Alrededor del 10% del platino es soluble en agua.

Con la nueva generación de catalizadores de tipo monolítico, los resultados de experimentos en plataforma de pruebas de motores con un catalizador de tres vías indican que la emisión total de platino es inferior por un factor de 100-1000 a la producida en catalizadores en pastilla. Con velocidades simuladas de 60, 100 y 140 km/h, se encontró que la emisión total de platino era de 3 a 39 ng/m^3 en los gases de escape, lo que corresponde a unos 2-39 ng por km recorrido. El diámetro aerodinámico medio de las partículas emitidas era de 4 a 9 μm en las distintas pruebas. Existen pruebas limitadas de que la mayor parte del platino emitido se encuentra en forma metálica o en partículas de superficie oxidada.

---

[a] La presente monografía se ocupa específicamente del platino y de ciertos compuestos del platino de importancia ocupacional y/o ambiental. No entra en el ámbito restringido de la serie de Criterios de Salud Ambiental el estudio pormenorizado de los efectos tóxicos del fármaco anticanceroso cisplatino y de sus análogos en el hombre y los animales, puesto que esas sustancias se usan principalmente como agentes terapéuticos. Además, sus propiedades tóxicas son excepcionales en comparación con las de otros compuestos de platino.

3. Transporte, distribución y transformación en el medio ambiente

Los metales del grupo del platino son escasos en el medio ambiente en comparación con otros elementos. En zonas muy industrializadas, pueden encontrarse cantidades elevadas de platino en los sedimentos fluviales. Se supone que la materia orgánica, por ejemplo los ácidos húmicos y fúlvicos, enlaza platino, proceso que tal vez se vea favorecido por condiciones apropiadas de pH y de potencial redox en el medio acuático.

En el suelo, la movilidad del platino depende del pH, el potencial redox, las concentraciones de cloruros en las aguas subterráneas y la forma en que se encuentra el platino en la roca primitiva. Se considera que el platino sólo será móvil en condiciones extremadamente ácidas o en aguas subterráneas con elevado contenido de cloro.

En los sistemas de ensayo *in vitro* se ha demostrado que algunos complejos de platino(IV), en presencia de platino(II), pueden sufrir metilación por la metilcobalamina bacteriana en condiciones abióticas.

4. Niveles medioambientales y exposición humana

Se dispone de muy pocos datos en cuanto a las concentraciones medioambientales debido a los reducidos niveles de platino en el medio ambiente y los problemas analíticos que ello acarrea.

Las concentraciones en muestras de aire obtenidas en las proximidades de autopistas en los Estados Unidos antes de la introducción del catalizador en los automóviles se encontraban por debajo del límite de detección de 0,05 pg/m^3. Algunos datos obtenidos recientemente en Alemania indican que en las cercanías de las carreteras las concentraciones de platino en el aire (muestras particu- ladas) varían entre < 1 pg/m^3 y 13 pg/m^3. En zonas rurales, las concentraciones se encontraban en un orden de magnitud similar (< 0,6 a 1,8 pg/m^3).

Las concentraciones de platino en el aire cercano a carreteras tras la introducción de los catalizadores de pastilla en los automóviles se han calculado basándose en modelos de dispersión y datos experimentales de emisión.

Resumen

Las concentraciones estimadas de platino en las carreteras y en las zonas próximas variaron entre 0,005 y 9 ng/m^3 para el platino total. Puesto que la emisión total de platino en un catalizador de tipo monolítico es inferior, probablemente por un factor de 100 a 1000, que en un catalizador de pastilla, las concentraciones de platino emitidas en ese tipo de catalizador se encontrarían en el margen de picogramos a femtogramos por m^3.

En el polvo depositado en las plantas de hoja ancha que bordean las carreteras en distintos lugares de California, se detectaron concentraciones de 37-680 µg por kg de peso seco. Aunque el número de muestras era limitado, los resultados indican que los catalizadores de automóviles liberan platino al medio ambiente próximo a las carreteras.

En experimentos en cámara vegetal, los cultivos herbáceos expuestos durante cuatro semanas a gases de escape ligeramente diluidos procedentes de un motor equipado con un catalizador de 3 vías (velocidad simulada: 100 km/h) no contenían platino con un límite de detección de 2 ng/g de peso seco.

Los estudios de las concentraciones de platino en sedimentos del Lago Michigan llevaron a la conclusión de que el platino se ha ido depositando en ellos durante los últimos 50 años a una velocidad uniforme. Las concentraciones en calas de sedimento de 1 a 20 cm variaron sólo entre 0,3 y 0,43 µg/kg de peso seco.

Mientras que no se han comunicado niveles de platino en aguas dulces, se han encontrado elevadas concentraciones (730 a 31 220 µg/kg de peso eco) en los sedimentos de un canal de corta sumamente contaminado en el río Rin (Alemania).

En muestras de *Pinus flexilis* se encontraron niveles de platino entre el límite de detección y 56 µg/kg (peso de ceniza). No obstante, el contenido de los suelos adyacentes se encontraba entre los mismos valores; estos datos limitados no indicaban tendencia alguna de acumulación.

En muestras aisladas de vegetales procedentes de un suelo ultrabásico, se encontraron niveles de platino de 100-830 µg/kg (peso seco).

En muestras de agua marina se han encontrado entre 37 y 332 pg/litro. En calas de sedimento obtenidas en el Pacífico oriental, las concentraciones de platino variaron entre 1,1 y 3 µg/kg (peso seco). La concentración más elevada (21,9 µg por kg) se encontró en sedimentos oceánicos alejados del litoral. En macroalgas marinas se han encontrado concentraciones de platino entre 0,08 y 0,32 µg/kg de peso seco.

En la población general se han medido niveles sanguíneos de platino de 0,1 a 2,8 µg/litro. En suero de trabajadores expuestos por su profesión, se han comunicado niveles de 150 a 440 µg por litro.

Se dispone de datos limitados sobre las concentraciones de platino en el lugar de trabajo. Es probable que los datos antiguos (0,9 a 1700 µg/m^3) no sean de fiar debido a las deficiencias del análisis. No obstante, esos datos permiten suponer que el nivel de exposición a sales de platino era superior al límite de exposición profesional de 2 µg/m^3 adoptado actualmente en la mayoría de los países. En recientes estudios realizados en los lugares de trabajo, se han medido concentraciones inferiores al límite de detección de 0,05 µg/m^3 o entre 0,08 y 0,1 µg/m^3.

5. Cinética y metabolismo

Tras una exposición única por inhalación (48 minutos) a distintas formas químicas del platino (5-8 mg/m^3), la mayor parte del ^{191}Pt inhalado fue rápidamente eliminado del organismo. A continuación se observó una fase más lenta de eliminación durante el resto del periodo posterior a la exposición. A los diez días de la exposición a ^{191}PtCl$_4$, ^{191}Pt(SO$_4$)$_2$, ^{191}PtO$_2$, y ^{191}Pt metálico, la retención total de ^{191}Pt por el organismo fue de aproximadamente 1, 5, 8 y 6%, respectivamente, de la carga inicial del organismo. La mayor parte del ^{191}Pt eliminado de los pulmones por mecanismos mucociliares e ingerido se excretó con las heces (semivida: 24 h). Una pequeña fracción del ^{191}Pt se detectó en la orina, lo que indica que la absorción en los pulmones y el tracto gastrointestinal fue muy reducida.

Resumen

En un estudio comparativo sobre el destino del $^{191}PtCl_4$ en ratas (25 μCi/animal) tras la exposición por distintas vías, la retención fue máxima tras la administración intravenosa, seguida por la exposición intratraqueal. La mínima se registró tras la administración oral. Puesto que sólo fue absorbida una cantidad minúscula del $^{191}PtCl_4$ administrada por vía oral, la mayoría atravesó el tracto gastrointestinal y se excretó con las heces. Al cabo de tres días, menos del 1% de la dosis inicial se detectó en todo el cuerpo. Tras la administración intravenosa, el ^{191}Pt se excretó en cantidades casi iguales tanto en las heces como en la orina. La eliminación fue más lenta que en el caso de la administración oral. A los tres días la retención en todo el organismo era de alrededor del 65%, y al cabo de 28 días aún era del 14% de la dosis inicial. A título de comparación, al cabo de periodos iguales alrededor del 22% y del 8%, respectivamente, quedaron retenidos por el organismo tras la administración intratraqueal.

Los principales lugares de depósito son el riñón, el hígado, el bazo y las glándulas suprarrenales. La elevada cantidad de ^{191}Pt encontrada en el riñón demuestra que una vez que el platino es absorbido, la mayor parte se acumula en él y se excreta en la orina. El nivel más bajo en el cerebro sugiere que los iones de platino atraviesan la barrera hematoencefálica sólo en grado limitado.

A diferencia de las sales hidrosolubles, el PtO_2, que es insoluble, sólo fue captado en cantidades insignificantes a pesar de que la sal se administró con la dieta en concentraciones sumamente elevadas, que representaron un consumo total de platino de 4308 mg por rata durante el periodo de cuatro semanas.

Tanto en el caso de las sales simples de platino como en el cisplatino, se ha determinado que existe un periodo de eliminación rápida seguido de una fase prolongada de eliminación durante el resto del periodo posterior a la exposición, y que no existen pruebas de que los perfiles de retención sean notablemente diferentes. No obstante, el cisplatino es sumamente estable en los fluidos extracelulares debido a que las elevadas concentraciones de cloruro suprimen la hidratación. Ello explica que se excrete principalmente en la forma no alterada. Su

excreción, a diferencia de la de las sales simples de
platino, tiene lugar principalmente con la orina.

6. Efectos en mamíferos de laboratorio y en sistemas de ensayo *in vitro*

La toxicidad aguda del platino depende principalmente
de la especie de platino. Los compuestos solubles son
mucho más tóxicos que los insolubles. Por ejemplo, la
toxicidad por vía oral en la rata (valores de la LD_{50}) disminuyo en el orden siguiente: $Na_2[PtCl_6]$ (25-50 mg/kg) >
$(NH_4)_2[PtCl_6]$ (195-200 mg/kg) > $PtCl_4$ (240 mg/kg) >
$Pt(SO_4)_2 \cdot 4H_2O$ (1010 mg/kg) > $PtCl_2$ (> 2000 mg/kg) >
PtO_2 (> 8000 mg/kg). No pudo alcularse la DL_{50} correspondiente a los dos últimos compuestos.

En las pruebas cutáneas realizadas en conejos albinos,
los compuestos PtO_2, $PtCl_2$, $K_2[PtCl_4]$, $[Pt(NO_2)_2(NH_3)_2]$,
$Pt(C_5H_7O_2)_2$ y *trans*-$[PtCl_2(NH_3)_2]$ se clasificaron como no
irritantes. Los compuestos $(NH_4)_2[PtCl_6]$, $(NH_4)_2[PtCl_4]$,
$Na_2[PtCl_6]$, $Na_2[Pt(OH)_6]$, $K_2[Pt(CN)_4]$, $[Pt(NH_3)_4]Cl_2$,
y *cis*-$[PtCl_2(NH_3)_2]$ resultaron irritantes en diversos grados.

En los ensayos de irritación ocular todos los compuestos de platino ensayados dieron resultados positivos.
El *trans*-$[PtCl_2(NH_3)_2]$ y el $(NH_4)_2[PtCl_4]$ resultaron ser
corrosivos.

Tras la inyección intravenosa de complejos de cloroplatino en cobayos y ratas, se observaron dificultades
respiratorias intensas, probablemente debidas a la
liberación analérgica de histamina. Esta liberación
inespecífica de histamina ha complicado la interpretación
de los estudios en animales y en el hombre en relación con
el diagnóstico de la sensibilización alérgica.

Tras la inyección subcutánea e intravenosa de
$Pt(SO_4)_2$ tres veces a la semana durante cuatro
semanas, no se observó inducción de un estado alérgico, de
acuerdo con las pruebas cutáneas (cobayos y conejos), la
transferencia pasiva y los ensayos en la almohadilla
plantar (ratones). La administración del complejo
platino-huevo-albúmina tampoco sensibilizó a los animales
de experimentación.

Resumen

No se consiguió sensibilizar a hembras de rata encapuchada de Lister con la sal libre de tetracloroplatinato de amonio, $(NH_4)_2[PtCl_4]$, aplicada por las vías intraperitoneal, intramuscular, intradérmica, subcutánea, intratraqueal y por la almohadilla plantar, con *Bordetella pertussis* como coadyuvante, de acuerdo con los resultados de la prueba cutánea directa, la prueba de anafilaxis cutánea pasiva o un ensayo de radioalergosorbencia (RAST). No obstante, se han comunicado resultados positivos de anafilaxis cutánea pasiva con conjugados platino-proteína.

En monos Cynomolgus (*Macaca fasicularis*) expuestos a hexacloroplatinato de sodio, $Na_2[PtCl_6]$, por inhalación exclusivamente nasal de una concentración de 200 $\mu g/m^3$ durante 4 horas al día, dos veces a la semana durante 12 semanas, se observaron insuficiencias pulmonares significativamente mayores que en los animales testigo. Con la exposición a hexacloroplatinato de amonio, $(NH_4)_2[PtCl_6]$, sólo la exposición simultánea a ozono (2000 $\mu g/m^3$) produjo hipersensibilidad cutánea e hiperreactividad pulmonar significativas.

En estudios de administración oral a machos de rata Sprague-Dawley, las sales $PtCl_4$ (182 mg/l de agua de bebida) y $Pt(SO_4)_2 \cdot 4H_2O$ (248 mg/litro) no ejercieron efecto alguno en la adquisición normal de peso durante el periodo de observación de 4 semanas. Al triplicar la concentración de platino, la adquisición de peso se redujo en un 20% sólo durante la primera semana, paralelamente a una disminución del 20% del consumo de alimento y agua.

Sólo se dispone de datos experimentales limitados sobre los efectos del platino en la reproducción, la embriotoxicidad y la teratogenicidad. El $Pt(SO_4)_2$ (200 mg Pt/kg) redujo el peso de las crías en ratones suizos ICR desde el día 8 al 45 después del parto. El principal efecto del $Na_2[PtCl_6]$ (20 mg Pt/kg) fue un nivel de actividad menor en las crías de madres expuestas el duodécimo día de gestación. Se considera que el alambre y las láminas de platino sólido son biológicamente inertes; los efectos adversos observados a raíz de la implantación en el útero de ratas y ratones se debieron probablemente a la presencia física de un objeto extraño.

Tras la administración intravenosa de $^{191}PtCl_4$ a ratas gestantes (25 μCi/animal) el día 18 de la gestación, una cantidad limitada del compuesto atravesó la barrera placentaria.

Se ha observado que varios compuestos de platino son mutagénicos en diversos sistemas bacterianos. En estudios comparativos, el cisplatino era varias veces más mutagénico que otras sales de platino ensayadas. En estudios realizados *in vitro* con células de mamífero (sistema CHO-HGPT), la actividad mutagénica relativa de los compuestos *cis*-$[PtCl_2(NH_3)_2]$, $K[PtCl_3(NH_3)]$ y $[Pt(NH_3)_3Cl]Cl$ fue 100:9:0,3. La mutagenicidad del $K_2[PtCl_4]$ y el *trans*-$[PtCl_2(NH_3)_2]$ era marginal, mientras que el $[Pt(NH_3)_4]Cl_2$ no era mutagénico. No se observó actividad mutagénica en los compuestos $K_2[PtCl_4]$ y $[Pt(NH_3)_4]Cl_2$, en el ensayo de letalidad recesiva ligada al sexo en *Drosophila melanogaster*, en un ensayo de micronúcleo de ratón ni en el ensayo en médula ósea de hámster chino.

Salvo en el caso del cisplatino, no se dispone de datos experimentales relativos a la carcinogenicidad del platino y sus compuestos. Existen prueba suficientes de la carcinogenicidad del cisplatino en los animales. No obstante, el cisplatino y sus análogos son excepcionales en comparación con otros compuestos de platino; ello se refleja en su mecanismo característico de actividad antitumoral. La constitución de enlaces cruzados entre las hebras de ADN, formados sólo por el isómero cis en determinada posición de la guanina, se considera el motivo de esa actividad antitumoral. Parece ser que la replicación del ADN es defectuosa en las células cancerosas, mientras que en las normales las lesiones causadas por el cisplatino en la guanina se reparan antes de la replicación.

7. Efectos en el ser humano

La exposición a sales de platino se limita principalmente al medio ocupacional, en particular a las refinerías de platino metálico y las plantas de fabricación de catalizadores.

Los princiaples compuestos responsables de la hipersensibilidad a las sales de platino son el ácido hexacloroplatínico, $H_2[PtCl_6]$, y algunas sales cloradas como el

Resumen

hexacloroplatinato de amonio, $(NH_4)_2[PtCl_6]$, el tetracloroplatinato de potasio, $K_2[PtCl_4]$, el hexacloroplatinato de potasio, $K_2[Pt_6]$ y el tetracloroplatinato de sodio, $Na_2[PtCl_4]$.[a] Los complejos en los que no hay ligandos halógenos coordinados al platino ("complejos no halogenados"), como el $K_2[Pt(NO_2)_4]$, $[Pt(NH_3)_4]Cl_2$ y $[Pt\{(NH_2)_2CS\}_4]Cl_2$, así como los complejos neutros como el cis-$[PtCl_2(NH_3)_2]$, no son alergénicos, puesto que probablemente no reaccionan con proteínas para formar un antígeno completo.

Entre los signos y síntomas de hipersensibilidad figuran urticaria, dermatitis de contacto de la piel, y trastornos respiratorios que pueden ir desde estornudos, disnea y cianosis a crisis graves de asma. El periodo de latencia desde el primer contacto con el platino hasta la aparición de los primeros síntomas varía desde unas pocas semanas a varios años. Una vez que la sensibilización está establecida, los síntomas tienden a empeorar durante el tiempo que los individuos sigan expuestos en el lugar de trabajo, pero suelen desaparecer al cesar la exposición. No obstante, si se produce una exposición prolongada después de la sensibilización, es posible que los individuos nunca queden completamente exentos de síntomas.

Aunque a partir de los datos disponibles no puede deducirse una relación inequívoca entre la concentración y el efecto, el riesgo de desarrollar sensibilidad a las sales de platino parece guardar relación con la intensidad de la exposición. El platino metálico parece no ser alergénico. A excepción de un solo caso comunicado de supuesta dermatitis de contacto provocada por un anillo "de platino", no se han notificado reacciones alérgicas.

---

[a] Se ha abandonado el término "platinosis" para describir las enfermedades relacionadas con las sales de platino, puesto que implica una enfermedad pulmonar fibrosante crónica del tipo de la silicosis. En su lugar, se han utilizado "alergia a las sales de platino", "alergia a los compuestos de platino que contienen ligandos halógenos reactivos" e "hipersensibilidad a sales de platino" (HSP), siendo preferible el último.

Las manifestaciones clínicas de la hipersensibilidad a las sales de platino reflejan una auténtica respuesta alérgica. El mecanismo parece ser una respuesta de tipo I (medida por la IgE). La posibilidad de que se desarrollen anticuerpos IgE a complejos de cloruro de platino en personas sensibles se ha supuesto basándose en los ensayos *in vivo* e *in vitro*. Se cree que las sales de platino de baja masa molecular relativa actúan como haptenos que se combinan con las proteínas séricas para formar el antígeno completo.

Las pruebas de punción cutánea con concentraciones diluidas de complejos solubles de platino parecen proporcionar indicadores biológicos de la alergenicidad reproducibles, fiables, razonablemente sensibles y sumamente específicos. Los compuestos utilizados para las pruebas periódicas de detección de alergias en los trabajadores expuestos son $(NH_4)_2[PtCl_6]$, $Na_2[PtCl_6]$ y $Na_2[PtCl_4]$. La sensibilidad y fiabilidad de las pruebas de punción cutánea no tienen igual en ninguno de los ensayos *in vitro* disponibles. En los inmunoensayos enzimáticos y las pruebas de radioalergosorbencia (RAST) se han encontrado anticuerpos IgE específicos de los complejos del cloruro de platino. Aunque se comunicó la existencia de correlación con los resultados de las pruebas de punción cutánea, la aplicabilidad de la prueba RAST para los exámenes de detección se puso en tela de juicio a causa de su falta de especificidad.

Se ha encontrado una reactividad cruzada limitada entre las sales de platino y de paladio en los ensayos cutáneos y el RAST. Las reacciones a los metales del grupo del platino distintos de éste sólo se han observado en individuos sensibles a las sales de platino.

El tabaquismo, la atopia y la hiperreactividad pulmonar inespecífica se han asociado a la hipersensibilidad a las sales de platino y pueden ser factores predisponentes.

En cuanto a la población general, no se dispone de bastantes datos sobre la situación real en materia de exposición en los países en los que se ha introducido el catalizador en los automóviles. Las concentraciones atmosféricas posibles, calculadas teniendo en cuenta algunos datos de emisión y modelos de dispersión, son inferiores por un factor al menos de 10 000 al límite de exposición ocupacional de 1 mg/m³ adoptado por algunos países para el platino metálico como polvo inhalable total. Puesto que el platino emitido se encuentra con toda probabilidad en forma metálica, el potencial de sensibilización de las emisiones de platino a partir de los catalizadores de los automóviles es probablemente muy bajo. Aunque parte del platino emitido fuera soluble y potencialmente alergénico, el margen de seguridad respecto del límite de exposición profesional para las sales de platino solubles (2 µg/m³) sería de al menos 2000.

En un estudio inmunológico preliminar, se ensayaron extractos de muestras particuladas procedentes de escapes de automóviles en tres voluntarios humanos mediante una prueba de punción cutánea. No se obtuvo respuesta positiva.

No se dispone de datos para evaluar el riesgo carcinogénico del platino ni de sus sales para el ser humano. En cuanto al cisplatino, las pruebas disponibles en materia de carcinogenicidad humana se consideran insuficientes.

8. Efectos en otros organismos en el laboratorio y sobre el terreno

Los complejos simples de platino tienen efectos bactericidas. El descubrimiento de que los complejos neutros como el cisplatino inhiben selectivamente la división celular sin reducir el crecimiento celular de diversidad de bacterias gram-positivas y especialmente gram-negativas ha llevado a su aplicación en medicina como agentes antitumorales.

El crecimiento y la cosecha del alga verde *Euglena gracilis* fueron inhibidos por el ácido hexacloroplatínico soluble (250, 500 y 750 µg/litro) en un "microcosmos" experimental. El cisplatino ocasionó clorosis y retraso del crecimiento en el jacinto acuático *Eichhornia crassipes* con una concentración de 2,5 mg/litro.

En el invertebrado *Daphnia magna*, la exposición durante tres semanas a ácido hexacloroplatínico ($H_2[PtCl_6]$), dio un valor de CL_{50} de 520 µg Pt por litro. Con concentraciones de 14 y 82 µg/litro, se observaron efectos en la reproducción que se manifestron en reducciones del 16 y el 50%, respectivamente, del número total de crías.

Tras la exposición a corto plazo al ácido tetracloroplatínico, $H_2[PtCl_4]$, en un bioensayo estático, los valores de la CL_{50} a las 24, 48 y 96 horas fueron de 15,5, 5,2 y 2,5 mg Pt/litro, respectivamente, en el salmón *Oncorhynchus kisutch*. Con 0,3 mg/litro se observaron efectos en a actividad natatoria general y el movimiento opercular. Con 0,3 mg/litro o más se observaron lesiones en las branquias y el órgano olfatorio. No se observaron efectos con concentraciones de 0,03 y 0,1 mg/litro.

Se han estudiado los efectos del platino en plantas terrestres; todos ellos se realizaron con cloruros de platino solubles. El ácido hexacloroplatínico inhibió el crecimiento de las plantas de judías y tomates en cultivo arenoso con concentraciones de 3×10^{-5} a 15×10^{-5} mol/kg (5,9-29,3 mg/kg). De nueve especies hortícolas cultivadas en solución hidropónica con tetracloruro de platino, $PtCl_4$ (0,057, 0,57 y 5,7 mg Pt/litro), se observaron reducciones significativas del peso seco en el tomate, el pimiento y las hojas de nabo, así como en las raíces de rábano con la concentración más elevada. Con esa concentración, los brotes y las hojas inmaduras de la mayoría de las especies sufrieron clorosis. En algunas de las especies, con niveles bajos de $PtCl_4$ se observó un efecto del estímulo del crecimiento. Además, se suprimió la transpiración con la concentración más elevada de platino, probablemente debido a una mayor resistencia de los estomas. También se observó estímulo del crecimiento con niveles reducidos de platino (0,5 mg Pt/litro), administrado en forma de tetracloroplatinato de potasio $K_2[PtCl_4]$, en plantones de la herbácea sudafricana *Setaria verticillata* cultivada en solución de nutrientes. Al cabo de dos semanas, las raíces más largas habían sufrido un aumento de longitud del 65%. Con la concentración más elevada que se aplicó, es decir 2,5 mg Pt/litro, se observaron efectos fitotóxicos en forma de retraso del crecimiento radicular y clorosis foliar.

www.ingramcontent.com/pod-product-compliance
Ingram Content Group UK Ltd.
Pitfield, Milton Keynes, MK11 3LW, UK
UKHW021311180426
11947UKWH00015B/1163